Praise for Driving the Autonomous Vehicle Revolution

"The authors make a powerful case that if regulators go beyond technology considerations and put human concerns at the center of decisions about AVs, we have a real chance to improve the quality of life in cities for millions of people."

— **M.J. Maynard,** CEO, Regional Transportation Commision of Southern Nevada

"Aimed at policymakers, urban planners, and concerned citizens, this in-depth analysis presents a powerful framework for how to handle the biggest paradigm shift since the Internet. An important book for anyone interested in how to ensure AVs have a positive impact."

— **Dick Alexander,** CEO, Transdev U.S.

"The authors bring their substantial expertise and insights to address the challenges facing cities as AV technology begins to impact communities. Their powerful framework will help city leaders, urban planners and policy makers make thoughtful decisions for the common good."

— **Michael Fleming,** CEO, Torc Robotics

"The best advice I have seen yet for city leaders and urban planners preparing for the onslaught of policy and political questions we will face as AVs are broadly adopted. A must-read for everyone concerned about improving how all segments of society live together and move around our cities going forward."

— **Paul Skoutelas,** President and CEO, American Public Transportation Association

"An important lesson in how auto culture came to dominate in the 1920s and beyond, leading to today's sprawling, often congested, and polluted cities. Will AVs follow this historic path or will decision makers and leaders employ this opportunity to focus on people and improve social equity and quality of life? A proactive approach to policy making and planning is needed to shape the future of cities."

— **Susan Shaheen,** Professor, Transportation Engineering and Co-Director, Transportation Sustainability Research Center, U.C. Berkeley

"After reading this pioneering book, we can no longer think of the AV without linking it to the 'TRUST' concept. It definitely guides us in our research on AVs and new mobilities."

— **Philippe Watteau,** CEO, VEDECOM Institute, France

ISBN Number - Paperback Edition: 978-1-7348858-0-4
ISBN Number - Digital Edition: 978-1-7348858-1-1

Library of Congress Control Number: 2020906486

Design by Dave Malone Design Group

Translated from French by Straker Translations

First Printing in the US: May 2020
First Printing in France: March 2019

Original title: *Piloter le véhicule autonome - au service de la ville*,
Paris, Descartes & Cie, March 2019

Publisher's Address:

Transdev North America, Inc.
720 East Butterfield Road
Suite 300
Lombard, IL. 60148

www.transdevna.com

Yann Leriche
Jean-Pierre Orfeuil

Driving the Autonomous Vehicle Revolution

Harnessing the Potential for our Cities

This book was written with the support and scientific contribution of the City on the Move Institute (IVM), a division of VEDECOM—the French Institute for the Energy Transition and a Public-Private Research Partnership dedicated to carbon-free, connected and autonomous vehicles and sustainable mobility.

Editorial Coordination by Mireille Apel-Muller

The authors would like to thank Mireille Apel-Muller, who brought this work to life with her questions, feedback and suggestions, the members of VEDECOM and Transdev for their clarification on technical points, Patrice Trapier for his careful proofreading and corrections, and Ruth Otte for her invaluable support in creating the English version of the book.

Any errors are our own.

Chicago and Paris, May 2020

To my wife Eva,
to my children, Elisabeth, Pierre, and Romain

— Yann Leriche

Table of Contents

"There's a lot of interest [in autonomous vehicles] and people tend to get distracted by this shiny new toy. Let's make sure that is the focus – creating the city that we want to have – and not looking at the technology as the be all and end all."

— Janette Sadik-Khan[1]

"Self driving cars will improve our cities. If they don't ruin them."

— Robin Chase[2]

1 Janette Sadik Khan is a former commissioner of the New York City Department of Transportation; she is the author of Streetfight: Handbook for an Urban Revolution, 2017, Viking, Penguin Books.

2 Robin Chase is the co-founder of Zipcar, the leading car-sharing company in the United States.

Introduction

The Autonomous Vehicle has its Visionaries

The Autonomous Vehicle has its Visionaries

The autonomous vehicle (AV) has already entered our lives, or at least our minds. No longer a far off, utopian ideal, the AV is now an imminent prospect, arriving in a matter of years. And of course, when it does arrive, it will be revolutionary.

This credo has been widely repeated by many observers, experts, and the media in general. It has almost become a matter of faith, judging by the lack of critical perspective in many predictions. Pronouncements by Silicon Valley giants, "GAFAM" or "NATU"[3], followed closely by the Chinese "BATX"[4], have swept away all objections in their path. And large consultancy firms have clambered over each other to deliver predictions that, despite their differences, uniformly announce a "disruptive innovation" looming large on the horizon.

To give just one example: PricewaterhouseCoopers (PwC)[5] estimates that by 2030, electric cars will make up 55% of total car sales, while shared vehicles will cover one-third of the total distances covered (which would likely lead to a 25% reduction in cars on the road in the United States and Europe), and most importantly, 40% of all mileage will be traveled in autonomous vehicles. Put simply, in just ten years' time, a new world will be upon us! The autonomous vehicle holds such power and fascination[6]

3 GAFAM: Google, Apple, Facebook, Amazon, and Microsoft; NATU: Netflix, Airbnb, Tesla, and Uber.

4 BATX: Baidu, Alibaba, Tencent, Xiaomi.

5 PwC, "EASCY, Five trends transforming the Automotive Industry," 2018, www.pwc.com. Accessed May 27, 2018.

6 This fascination is nothing new. As the October 12, 2018 edition of the French financial newspaper *Les Échos* recalls, the front page of its first daily issue (in the summer of 1928) bore the headline "The driverless car," accompanied by a photo.

that it has caused some of the most prestigious consultancy firms, PwC included, to take leave of their usual critical faculties. Even car manufacturers have been carried along, at least temporarily, by this wave of euphoria.[7] Not so long ago, they were reluctant to jump onto the bandwagon of AV hype, out of concern that their customers would be deprived of the joy of driving. The conversation then was: driving assistance, yes; self-driving cars, absolutely not! Having done their market research, manufacturers knew that drivers were still mostly suspicious of the AV, some hostile, even in the United States, and that they were not inclined to spend more to own a car that drove itself.

Those manufacturers therefore kept a cautious eye on the movement, continuing to gradually introduce advanced driver-assisted features as they had been doing since the 1970s.[8] The science journalist Burkhard Bilger reported that at the New York Auto Show of 2013, the self-driving car was still a taboo subject.[9] Jeep and Mercedes presented advanced features, taking care to avoid the term "self-driving," while a Ford executive said, "Our view of the future is that the driver remains in control of the vehicle. He is the captain of the ship."

Then, from 2010 to the mid-teens, the manufacturers too began to get caught up in the groundswell, with all the zeal of the newly converted. They stepped up the pace, announcing a raft of upcoming release dates for their autonomous vehicles. Such was the hype around self-driving cars that, in a matter of just a few years, one of the most powerful industries in the world has been forced to stop and reconsider its position.

Unless, that is, this dramatic conversion was nothing more than a tactical or public relations move—we shall see. Then it was the major world powers' turn to hop on board. Almost all of them were signatories to the 1968 Vienna Convention on Road Traffic, which states: "Every moving vehicle [...] shall have a driver," and "Every driver shall at all times be able to control his vehicle." They amended this text in 2016 under the auspices of the United Nations Economic Commission for Europe (UNECE)[10] in

---

7 However, in recent months, the CEOs of some car manufacturers have attempted to rein in optimism, indicating that it will take time before we see the arrival of the "fully autonomous" vehicle.

8 Power steering, anti-lock braking systems (ABS), cruise control, electronic stability control (ESC), valet parking, anti-collision radar system, emergency brake-assist, etc.

9 Burkhard Bilger, "Auto Correct: Has the self-driving car at last arrived?" *The New Yorker*, November 25, 2013.

10 While most developed countries have ratified this convention, the United States, Great Britain, Spain, China, and the Vatican have not.

order to better mobilize the talent, energy, and research funds necessary for self-driving cars. Everyone wants to be in the running, or even better, to be first to the finish line. Self-driving cars are lauded with every virtue and have almost miraculously avoided the condemnation of the anti-car sentiment that has prevailed in several European countries on and off for the last twenty-five years.

Tech has Two Schools of Thought

How was such a reversal possible? We could summarize the situation with a sound bite: like Caesar's "*veni, vidi, vici*" ("I came, I saw, I conquered"), Google took similar action. With considerable resources and a complete mastery of software development, especially in the field of artificial intelligence, Google and a few other Internet giants invested in this field of transport, and emerged victorious, having fundamentally altered the automobile landscape, at least on the surface.[11] Anthony Levandowski, who led the "Google Car" project until 2016, described the company's success as follows[12]: "They want to make cars that make drivers better. We want to make cars that are better than drivers." Sergey Brin, the co-founder of Google, drove the point home: "We are just on such a different planet." There it is: a paradigm shift has occurred. It remains to be seen whether this "different planet" is going to be viable, and under what circumstances. This is where a second paradigm shift comes in.

Brin went on to lament the enormous amount of land taken up by parking lots and the proliferation of multi-lane highways, when private automobiles are only used for one or two hours a day. In Brin's view, if cars could drive themselves, there would be no need to own them. A fleet of self-driving cars could operate as a personalized transportation service, picking up passengers and dropping them off wherever they wanted, with the added bonus that the vehicles would be much cheaper than taxis and more flexible than buses. Fewer cars would be needed, which would lead to less congestion on the roads. Streets would be cleared, highways would shrink, and parking lots would become public parks. An efficient yet "frugal" city:

---

11 Let's not forget the words of storytelling expert Annette Simmons: "People don't want more information. They want faith [...]. It is faith that moves mountains, not facts." The Story Factor, Basic Books, 2002.
12 Burkhard Bilger, *op. cit.*

haven't we all dreamed of such a thing? Well, Google is offering it to you on a digital platter— and one bite would presumably take us to paradise!

Credible or not, this vision of improving the urban environment using autonomous vehicles is now shared by almost all of the consulting firms, along with many of the major players in the field of transportation. One example is an initiative led by the Boston Consulting Group (BCG) in association with all the main transport operators in the greater Paris region of Île-de-France.

The reinvention of mobility is inevitable: technological revolutions and new disruptive modes of transportation are going to profoundly change the way people travel, the vehicles they use and regional environments as a whole. This reinvention must be built around three technological disruptions (connected mobility, zero emissions and autonomy) and three disruptions in user behavior (on-demand, shared and co-modal mobility). These changes are already underway, and technology will not be the main obstacle to their development.[13]

However, not all of the stars of Silicon Valley share this vision of the autonomous vehicle as a catalyst for declining car use, and therefore the savior of a planet in the midst of global warming. Elon Musk envisages a dual system at first: selling his vehicles to people who will own them, while at the same time creating a "Tesla network" that will allow users to share their vehicles with friends or other highly rated (five-star) drivers, if and when they want. Unlike Brin, Musk believes that car use will become so inexpensive[14] that there will be more traffic than before, leading to more traffic jams, which will justify the construction of an underground tunnel network, a project that is being developed by another of Musk's companies

13 Boston Consulting Group, "Mobility Nation: réinventer la mobilité urbaine et périurbaine à l'horizon 2030," 2017, www.mobility-nation.com. It should be noted that all of these terms used are technical and are also based on societal meanings: "zero emissions" means "no emissions from use," ignoring emissions related to power generation or battery manufacturing and vehicle particulate emissions. "Connected" refers to modernity, but the nature of that connection remains unclear. "Shared" refers to car sharing rather than the use of public transportation, which is also a form of sharing. "Autonomous" refers to the high expectations of people today (of not being told how to behave by anyone), while vehicle autonomy means submitting to algorithms and giving up control to the vehicle. "On demand" refers to the personalization of transportation services, which is well in keeping with the expectations of people today, but overlooks the fact that the privately-owned car meets the demand of "when I want and where I want." The "co" in "co-modality" refers to a concept of sharing (e.g., coworking), whereas the concept of intermodality refers to constraints (changes in modes of transportation and connections). All of this semantic guesswork is indispensable if we are to build a convincing argument in favor of disruption. However, they may also become obstacles to implementing solutions.

14 i.e., less expensive and much more comfortable than public transit, including the subway.

that specializes in tunnel construction.[15]

It is reasonable to assume that the difference in vision between these two schools of thought could be based on opposing cultural frames of reference. Moscow native Sergey Brin was steeped in a collectivist culture, while in Praetoria, South Africa, Musk experienced a world where sharing a trip with strangers was unimaginable.

In Sergey Brin's view, a car will be something that is shared with other people.[16] For Musk, when you need to travel, you will call *your* car, which will also be able to pick up a pizza on its own, and it will drop you at your destination, before finding a parking lot on the outskirts of town, where parking is much cheaper. Owners will also be able to lend or rent out their vehicles, but not to just anyone. It will be a "cocoon city"—a contactless city, which is almost the exact opposite of the frugal city. In both cases, the car drives itself, but the underlying circumstances are entirely different.

This is not to say that these two visions are the only ones possible, but they provide enough contrast to show us that the vision of a self-driving car single-handedly solving all mobility problems is a fantasy. Like most innovations, the autonomous vehicle could have both positive and negative effects, and many of its assumed vices and virtues will depend on the systems into which it is introduced.

A Disruptive Innovation in Search of a New Breath

Late 2017 and early 2018 marked a turning point in the recent history of the autonomous vehicle, as if the initial wave of euphoria was beginning to fade away. Executives of major German car companies decided to inform the Brussels Commission that it would take some time to produce "fully

---

15 Fred Lambert, "Tesla Network: Elon Musk elaborates on autonomous ride-sharing, says will eventually be cheaper than public transport," *electrek*, May 1, 2017, electrek.co/2017/05/01/tesla-network-elon-musk-autonomous-ride-sharing-vision.
16 A situation not to Elon Musk's liking: "On public transport, there's a bunch of random strangers, one of who[m] might be a serial killer...." Remarks reported by Aarian Marshall in the magazine *Wired*, "Elon Musk Reveals His Awkward Dislike of Mass Transit," December 14, 2017, www.wired.com/story/elon-musk-awkward-dislike-mass-transit.

autonomous" Level 5 vehicles (see Chapter 5 for definitions of the five levels of autonomy). Toyota's head of research was similarly cautious. The auto industry gave off mixed messages at the 2018 Paris Motor Show, demonstrating that car manufacturers are able to keep one foot in the old world and the other in the new. The poster for the event showed two hands holding an invisible steering wheel, and the trade show had dedicated areas such as "Mondial.Tech" and the "Salon de la Mobilité" (Mobility Showcase), in keeping with the spirit of disruption. At the same time, the event also hosted a vintage car parade in the center of Paris, along with an exhibition on legendary roads that evoked the joys of driving. In the introduction to *Histoire* magazine's Paris Motor Show special, the organizer signed off with the words "take the wheel."

The report submitted by Anne-Marie Idrac, Senior Head of the National Strategy for the Development of Autonomous Vehicles in France, advises a degree of caution: "The objective here is to facilitate the emergence, then deployment, of innovative automation technology, and to support, at its own pace, the technological progress proposed by manufacturers, while ensuring a high level of safety: without adequate levels of safety, there will be no user confidence in new mobility services."[17] It should be noted that the former French Minister of State for Transport was referring to innovative technology in terms of automated driving capabilities, not fully autonomous vehicles.

A similarly cautious approach has also been called for in the United States, after the fatal collision of an Uber vehicle with a pedestrian, in an incident that sent shockwaves around the world. At last, politicians in Europe and North America are becoming aware of the need for oversight and regulation of the enormous financial and ideological clout of the tech giants, given their potential to massively impact the way in which our societies move forward. So we now see Uber, a company that appeared out of the blue and set up shop everywhere without asking anyone for their opinion, becoming subject to regulation and having to negotiate its position in the local mobility ecosystems of cities.[18]

---

17 Anne-Marie Idrac, "Développement des véhicules autonomes. Orientations stratégiques pour l'action publique," May 2018.

18 Alexis C. Madrigal, "Uber and Lyft Are Uniting Against Their Common Foe: The Private Car," *Citylab*, June 7, 2018, www.citylab.com/transportation/2018/06/will-uber-and-lyft-become-different-things/562319.

An Opportune Moment for Reflection

It is always a mistake to be right too soon[19] or try to persuade when there is panic in the air. This was the case during 2012–2015, when car manufacturers began to fear they would be swallowed up, or even steamrolled by, the Californian tech giants. So they shared their worries with their respective governments. At the time, anyone attempting to start a reasoned debate about the future of autonomous vehicles—let alone the pros and cons of the technology for people's lives and the urban environment—would have been permanently labeled a dinosaur. At that time, the autonomous vehicle was rapidly climbing the Gartner[20] Hype Cycle[21] (see Figure 1 on the following page).

This graph describes the change in interest in a new technology in five phases:

1) **The Innovation Trigger**, when interest is first being generated;

2) **The Peak of Inflated Expectations**, which corresponds to the highest point on the curve;

3) **The Trough of Disillusionment**, when the product fails to live up to overly enthusiastic expectations;

4) **The Slope of Enlightenment**, when the second-generation products meet revised expectations; and

5) **The Plateau of Productivity**, when the new technology finally becomes well established.

19 A Ford executive, who had publicly stated that an autonomous Ford car was not going to come about anytime soon, immediately lost his job.

20 Gartner Inc. is a U.S.-based advanced technology consultancy and research firm.

21 Hype: media hype, hyper-mediatization

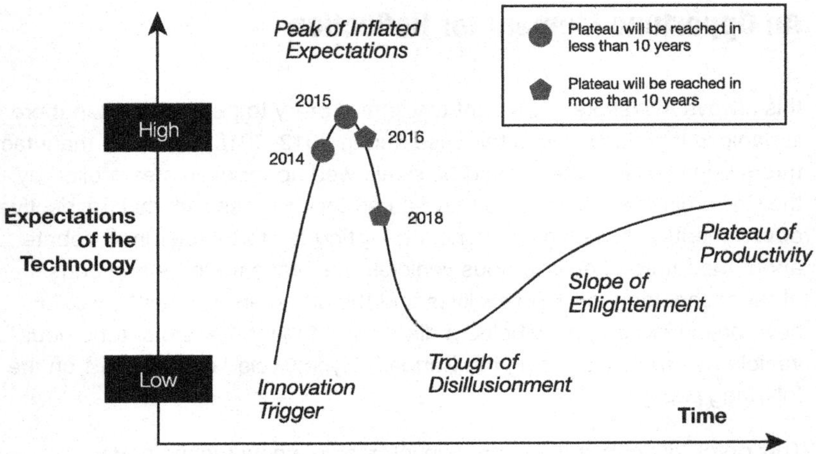

Figure 1. Position of the Autonomous Vehicle on the Gartner Hype Cycle Curve

Source: Jean-Pierre Orfeuil, Yann Leriche, excerpted from the annual publications of Gartner, Inc.

Autonomous vehicles reached the peak of the curve in 2015–2016, and Gartner now places Level 4 vehicles (those that are self-driving in some situations) on the downward slope, leading toward the Trough of Disillusionment. As the rest of the cycle shows, this does not spell the end of the autonomous vehicle; it indicates that the AV will rise again, transformed, to meet adjusted expectations. By betting on its future existence, we avoid the risk of being overwhelmed by a technological wave that is well beyond our control. In contrast to the swooning proponents of the autonomous vehicle, it is necessary, indeed urgent, that we take a little time and a big step back. If we want to adjust the inflated expectations, it would be useful to revisit the past impacts of new methods of transportation on city life, to discuss the future of the autonomous vehicle and how it may change our cities and lives depending on how it is used, and to invest in new forms of usage that its creators may not even have imagined. Our critically analytical approach does not overlook the fact that the autonomous vehicle has a future, but it shows that a disruption of this kind will need to be guided, specifically by public authorities, if it is to be successful.

Moving Beyond the Technological Illusion

In this book, we will be both taking our time and moving quickly.

We will take our time because a major evolution in our human journey deserves to be discussed and debated with full knowledge of the facts. We cannot claim to be in a knowledge economy and engage in such ground-breaking disruptions while we are in panic mode or under the influence of sensation-driven news cycles.

And we will take our time because the autonomous vehicle is still an unfinished product, not a refrigerator that can be simply plugged in and left to do its job of cooling and freezing untouched for ten years. The self-driving car is a constantly evolving, ever-developing product.[22]

While we need to take our time, we also need to move quickly: in our collective imagination, the autonomous vehicle has already become a reality. Technology moves fast, and real-world testing has already begun. We must move quickly to imagine the ways in which the AV should be introduced, knowing that its benefits can only be harnessed if it is integrated into systems that we, the citizens and urban stakeholders, must build and sustain.

We believe—and this is a critical point at odds with the current ideology—no mobility-related technology can be properly assessed without first being seen as a "social object" that affects society, human beings, and life in cities, even before being brought into service.

More specifically, we assume that as technology advances, autonomous vehicles will be available in the foreseeable future, and that along the way different stakeholders[23] will be interested in the potential of different types of vehicles, in different regions or transportation networks. Logically, then, these vehicles will play an important role in people's mobility, and therefore they deserve serious thought.

22 This issue will be explored in Chapter 1.

23 e.g., private individuals, businesses, authorities, and transportation companies, etc.

The AV: a Social Object

Our aim is to think through how these new objects will be introduced into our lives and our cities, and to suggest a framework for assessing the potential benefits and risks these vehicles will pose for individuals and the community, in both the short and the long term.

For that reason, we have chosen to start by discussing the AV's potential uses. Taking this approach in itself triggers an avalanche of questions:

- Which vehicles in particular are we talking about? Cars, minibuses, vans, trucks, other vehicles? The vehicles that exist today or those that will be invented in the future?
- What level of autonomy are we referring to?
- Who will be responsible for them? People, for private use? Groups of people, for community use? Transport operators, with vehicles that are accessible to all, operating car-sharing, taxi, or on-demand transportation services, which may be shared to differing degrees, as a regular public transportation service or for delivery?
- What will be the travel experience for a child traveling alone to school? A young woman sharing a journey with two drunk young men?[24] A business person rushing to an appointment?
- What are the objectives? To transport goods or people as efficiently as possible from one point to another, freeing them from the task of driving from point A to point B? To enable the development of new mobility solutions? To drive us to a point where a public transport service is waiting for us?
- Will governments be able to define their role and keep a tight rein on these systems? Where will intelligence be at this point? Who will authenticate it? Who will act as a trusted third party? What rules will these vehicles be subject to, for example, with regard to pedestrians?

---

24 Alcohol consumption is currently one of the main reasons why people use on-demand transportation in the United States.

- What personal data will be shared, intentionally or unintentionally, with the operators of such systems, and what will they be able to do with that data? Will these vehicles be confined to particular spaces or types of networks, or will they be used universally?

The introduction of autonomous vehicles raises a lot of questions. We will need a key to enter this new realm. For us, this key will be: TRUST.

TRUST: a Key for Managing the Arrival of the Autonomous Vehicle

The story of the AV remains to be written. Long before the autonomous vehicle enters our day-to-day lives, many developments are still to be expected and many heated debates will take place. Our goal is to shed light on the issues involved.

We will organize them using the acronym TRUST[25], which defines the five dimensions to be taken into consideration when examining the arrival of autonomous vehicles:

1) Technology

2) Regulations

3) Uses

4) Systems and services

5) Territories/regions

More than just a play on words, TRUST is a rigorous, analytical framework that provides a methodology for understanding the interactions between complex phenomena. It is a tool that can be used to build a model of the

25 This acronym is also a reminder that confidence in the proposed technologies (on the part of politicians, citizens who use public spaces, and potential users) and the impacts caused by the existence of these vehicles will be a decisive factor in how widespread these vehicles become.

changes currently underway and a key to understanding them. Each of the five links in the "TRUST" chain is important. If we lose sight of the chain as a whole in its logical order, or play down any one of its links, we weaken the system and ultimately open it up to failure, just as the current unilateral analyses of the AV are at risk of doing.

Before starting out on our analysis using the TRUST framework, we will briefly define what we mean by "autonomous vehicle" (Chapter 1). We will go on to present the historical relationship between mobility and the city (Chapter 2). Then we will compare the vision for autonomous vehicles and the current mobility systems in which they will be integrated, with the related pros and cons (Chapter 3). To be as comprehensive as possible, we will discuss many different points of view put forward by various players in the AV industry, particularly in the United States and Europe (Chapter 4), before taking five chapters to examine the challenges involved in the arrival of autonomous vehicles from the perspective of the five links in our TRUST framework.

Chapter 1

The Autonomous Vehicle: What is it? Where are we with it Today?

Although it may seem a straightforward concept at first glance (the autonomous vehicle is a vehicle that travels without a driver), the reality is a bit more complex. Autonomy is a vague notion that has no single clear definition, but we will find our bearings with the help of an official five-level scale and several possible avenues toward vehicle autonomy.

Inherent Problems of Terminology

The English language has two terms to refer to the vehicles we are discussing in this book: "self-driving" and "autonomous." This duality implies a bias that must be dispelled from the outset. Autonomy can be interpreted as a quality possessed by an individual who is able to live as he or she sees fit. For example, autonomy can describe the state of a young adult who is free of the norms inherited from their parents and who may find it useful, enjoyable, or even necessary to distance themselves from convention. In short, the autonomous individual can act in a way that is self-determined and unpredictable.

Thankfully, the so-called autonomous vehicle is, in theory, the exact opposite of this definition. Nothing is more predictable than an autonomous vehicle. Not only will it always react the same way in two identical situations, but all AVs are expected to make the same decisions, at least based on the same informational input. The term "autonomy" refers to the fact that the vehicle does not need a driver (or even passengers for that matter) or any external support to make its journey (e.g., from connected infrastructure). In this respect, autonomy marks a break with decades of research and testing on automated highways, where "intelligent" infrastructure guides the vehicles that travel on them.

Despite the highly homogeneous and predictable nature of AV driving, there is no strict, unambiguous definition of the capabilities of the autonomous vehicle in the real world. To illustrate, we can go back to our comparison with the young adult. An autonomous young person will be less autonomous once they join an organized group or setting, for example, when doing military service, when working on a production line, or when in prison. On the other hand, a person can (depending on good behavior) still be autonomous to a certain extent within the space of their prison. In much the same way, a vehicle, too, can only be recognized as autonomous in certain contexts.[26]

Ultimately, there is no one definition, set in stone, for the AV, but rather a commonly accepted scale of autonomy. This scale might seem to suggest that the definition of the autonomous vehicle is based on a purely progressive approach, advancing from Level 1 to Level 2, and so on. But again, this is not the case. Further on, we will outline the two strategies currently being implemented by the main players: a progressive approach for "all-purpose" vehicles, and an approach that is directly targeted toward the higher end of the scale, for use in specific predetermined contexts.

In an attempt to get a clearer picture of the issue and overcome the difficulties of settling on a non-anecdotal definition, we will examine autonomy from the perspective of technological progress, followed by transportation systems, and finally regulatory environments.

Five Steps Toward Autonomy: An Official Scale

There is a universally accepted scale of autonomy, which combines the legal and technical characteristics relating to driver responsibility. The U.S. scale has been defined by SAE International[27], while the international scale has been defined by the IOMVM.[28] For the most part they converge: both have five levels, in addition to Level 0, where the driver does everything but can be

26 Paradoxically, it's in a more restricted environment that the autonomous vehicle is able to make the best use of its autonomous capabilities, much like a piece of farm machinery that runs on strict grid or circular patterns in a field in the Midwest.

27 Society of Automotive Engineers

28 The International Organization of Motor Vehicle Manufacturers

given automated alerts, such as those from a reverse parking radar system.

Level 1: the driver is assisted

The onboard computer can control the vehicle's speed OR steering, but not both at the same time. The driver then keeps control of the other functions and full control of the vehicle, i.e., the driver's commands override those of the machine, even the computer-controlled function. Adaptive cruise control is a good example of a Level 1 "autonomous" driving experience, where the car slows down to maintain a safe distance depending on the car ahead. However, the driver remains fully responsible for the vehicle, including the driver-assist function (steering or speed).

Level 2: the vehicle has an advanced driver-assistance system and the driver becomes a supervisor

At this level, the vehicle can control speed AND steering in certain circumstances. An example of this would be parking systems, where the onboard computer temporarily takes over the task of driving. However, the driver must still pay close attention, supervise the vehicle's maneuvers, and be ready to take back control immediately on their own initiative. The driver's "commands" will always override those of the machine. As with Level 1, the driver remains fully responsible: they have a duty of supervision. In the coming years, vehicles produced at the high end of the market are expected to continue to gradually incorporate features from these two levels of autonomy, since they do not involve any regulatory changes.

Level 3: the vehicle is equipped with a situation-dependent autonomous driving system

The driver fully delegates the task of driving in predefined situations, but there must be a driver on board, and the driver must be able to instantly take back control as soon as the system requests it. Aside from these situations, the driver simply drives. A typical example would be low-speed driving in traffic jams. In a gridlock situation, the driver could read the newspaper, but the onboard computer could ask the driver to take back the controls as soon as the situation improves (even if they haven't finished reading their article).

Level 4: a driver is no longer needed in certain situations

The vehicle is able to travel without its owner in specific circumstances that it can define for itself. For example, a car can park itself in a parking lot. It will also be able to come back to the driver when needed. The driver no longer has to be on board and does not have to worry about what the car is doing. This level is key for transportation services. These services will be deployed within well-defined regions and networks, under specific conditions, and the fact that a driver is no longer needed may represent an essential part of the economic model for these vehicles.

Level 5: the vehicle is fully autonomous in all situations

At this level, the traveler simply inputs the destination and the onboard computer takes control of all of the vehicle's functions. The vehicle is programmed to never have to ask the passenger (if there even is one) to take control, in any circumstances. It is probable that Level 5 autonomous vehicles will have no steering wheels or pedals.

Two Possible Avenues Toward Autonomy

The five-level SAE and IOMVM scales suggest that the progression toward autonomy is linear, i.e., Level 5 comes after Level 4, which in turn comes after Level 3, etc. However, this is not the case. Two aspects must be considered: 1) the vehicle's autonomous capacity in relation to its driver and 2) the complexity of the use cases.[29]

There are several ways to get from Level 0 (where the driver does everything, in all situations, without the help of a computer) to Level 5 (i.e., absolute autonomy). One of these ways, which we will call the "horizontal avenue" (see Figure 2), consists of incrementally equipping vehicles. These vehicles are authorized to drive in all use cases, even the most complex, first with driver-assist technology, and subsequently with self-driving functionality. Historically, this has been the route taken by traditional car

---

29 e.g., the layout of roads, level of traffic, or weather conditions. There are many different factors, each of which will need to be isolated on closer analysis.

manufacturers who, year after year, have sold models with continuously improved features, without limiting their use to certain locations or placing conditions on their use. The second way, the "vertical avenue," has been favored by manufacturers of autonomous shuttles[30] in particular (see Vehicle B in Figure 2). These vehicles are completely autonomous, requiring no human intervention (including no steering wheels or pedals), but their operation is limited to simple use cases. The technologies developed by these manufacturers aim to increase the complexity of the environments in which they are able to operate.

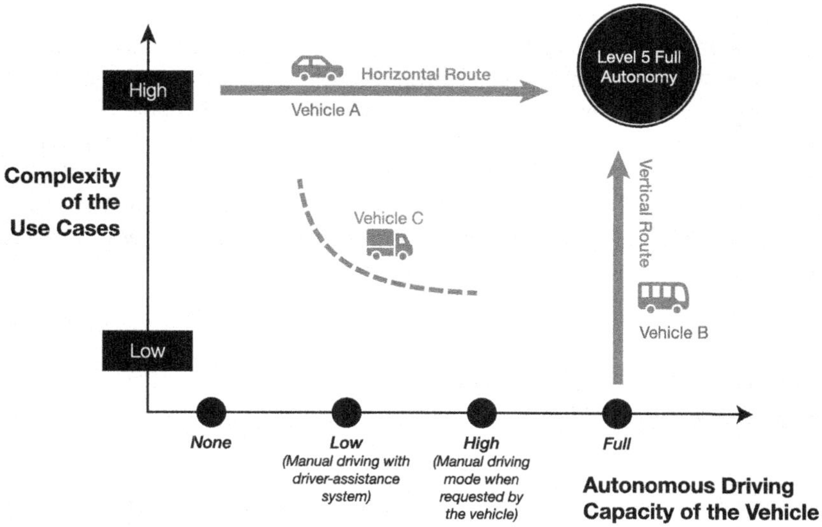

Figure 2. Two Avenues Toward Autonomy

Source: Yann Leriche, Jean-Pierre Orfeuil

Beyond technical considerations, these two avenues also represent two differing philosophies of innovation. The first assumes that space (and the use of space) will adapt to a new technology that will be so powerful it will eliminate any barriers. In the second, space constraints are used as a starting point, and the aim is to make changes to the way services are provided. Other types of vehicles, both those that currently exist and those that will

30 These are currently manufactured by companies such as Navya, EasyMile, Optimus Ride, and Local Motors.

exist in the future, are shown in Figure 2 in an intermediate position. They are not necessarily represented by a single point on the diagram, but can be identified by multiple positions (see Vehicle C). Indeed, a vehicle may be able to drive completely independently on highways and in city centers where its use has been authorized, while it may only offer driver assistance in other use cases.

These different positions, which can be occupied by the same vehicle, demonstrate why discussions about levels of vehicle autonomy are usually extremely vague: the precise subject of the conversation is not always clear. As for the question "When will autonomous vehicles be available?", there are as many answers as there are use cases, even if we agree that "autonomous vehicle" means a vehicle without a steering wheel or pedals. It is also possible, in fact probable, that one day we will all travel in vehicles that are capable of fully autonomous driving, without ever reaching the absolute autonomy of Level 5. In this case, each vehicle would be approved for certain uses (e.g., traveling in a given area) but would not be approved for universal use. This is the case today for urban transportation systems in many countries, which are subject to certification based on local operating factors.

Where are we Today from a Vehicle Perspective?

Beyond the objective difficulty of classifying the performance of an autonomous vehicle or transportation system (not only the vehicle itself, but also what makes it work: a remote control center, secure communications, etc.), many tests and demonstrations involve vehicles that are and will remain prototypes. This raises a question about the credibility of the announcements surrounding such events, particularly when the press releases for these prototypes fail to provide any information on the tests, costs, and production capabilities involved. As the U.S. academic Daniel Boorstin said in 1962, well before the autonomous vehicle existed, "We live in a world of pseudo-events and quasi-information, in which the air is saturated with statements that are neither true nor false but merely credible."[31]

31 Daniel J. Boorstin, *The Image: A Guide to Pseudo-Events in America*, Vintage Books, New York, 1962.

The only certainty we can have is that all the main players involved have an interest in effective communication, because "in the post-industrial society... the challenge is not in the social use of technology, but the mass production and dissemination of representations, information and languages... The production of symbolic goods has taken on the central role once occupied by material goods in the industrial society."[32] And it is precisely this certainty that then leads us on to doubt.

Independent analyses can help to reduce this uncertainty to some extent. The market research and consulting firm Navigant Research[33] ranks the major players in autonomous vehicle manufacturing annually according to objective criteria. Based on ten criteria reflecting both the strategy of the manufacturers and their implementation capacity, it divides companies into three groups: leaders, contenders, and challengers. Eight companies are labeled leaders in the latest report, and two of them are at the forefront: General Motors and Waymo, a subsidiary of Google. The other members of the leaders' club are Daimler, BMW, Volkswagen Group, Ford, the Renault-Nissan alliance, and Aptiv.[34] Now, if we look to the other end of the scale (the challenger category), we see Apple, Tesla, and Uber.

The Navigant report is an interesting study, both for its professional content and for the debates it generates[35], both subjects that we will discuss later in this book. Although Navigant cautions that its ranking is only valid at a given moment (as evidenced by the many changes made from one year to the next), its ten evaluation criteria have been subject to a number of consistent criticisms. These ten criteria are divided into two groups:

Group 1: Strategy

- **Vision**
- **Marketing strategy**
- **Partners**
- **Production strategy**
- **Technology**

Group 2: Implementation

- **Sales, marketing, and distribution**
- **Product performance**
- **Product quality and reliability**
- **Product portfolio**
- **Financial strength**

---

32 Alain Touraine, *Critique de la modernité*, Fayard, Paris, 1992.

33 Navigant Research, "Navigant Research Leaderboard: Automated Driving Vehicles," navigantresearch.com, 2018.

34 A company that resulted from the breakup of Delphi, one of the leading auto parts makers in the world, into two separate companies: Aptiv and Delphi Technologies.

35 See, in particular, the article by Alex Roy, "Who is Really #1 in Self-Driving Cars? Navigant's controversial report doesn't use enough criteria," *The Drive*, April 7, 2017, www.thedrive.com/opinion/9024/who-is-really-1-in-self-driving-cars-you-wouldnt-know-it-from-navigants-controversial-report.

Focusing on the three main criticisms that have been leveled at this list, the first is that it is based on the implied assumption that the business models of tomorrow's automotive industry will be comparable to those of today, which revolve around product sales. However, autonomy is "disrupting" the transportation industry and significant changes in use and value chains are anticipated, as glimpsed in the visions of Elon Musk and Sergey Brin mentioned in the introduction.

The second criticism relates to the weighting of technology in the evaluation, which only counts for one tenth of the total rating, when the creation of an autonomous vehicle is very much a software-related endeavor. That is the Californian vision, at any rate, and the counter argument from Detroit should not be disregarded: manufacturing a mass-produced vehicle requires specific skills that are not easily acquired, as Tesla discovered with its Model 3.[36]

The third criticism highlights the absence of two key criteria from the list: community relations and data. Transportation is a key factor in the quality and appeal of a city or region, and brings with it many safety issues, so local government officials are unlikely to lose interest. According to the former mayor of New York City, Michael Bloomberg, "Autonomous vehicles have the potential to complement the work cities are doing—so long as people, not cars, drive the agenda."[37] In addition, accumulated data feed artificial intelligence software, and the data's intrinsic quality is of little interest if the scope of its learning terrain is limited. Regarding this criterion, Tesla is certainly not trailing behind the other companies, as its fleet of connected vehicles is already in circulation, gathering huge amounts of data, mile after mile.

Ultimately, who is leading the field? In our view, we would place two companies at the top of the leaderboard: Waymo and Baidu. The U.S. company, Waymo, was launched first; it has considerable financial and technological resources and has already shown what it is capable of. The Chinese company, Baidu, with its Apollo project, has created a complete,

---

36 Leon Poultney, "Hey Tesla, how hard can it be to actually make a car?" *Wired*, April 22, 2018, www.wired.co.uk/article/tesla-model-3-production-stock-problems-engineering.
37 Michael Bloomberg, "Autonomous vehicles have the potential to complement the work cities are doing – so long as people, not cars, drive the agenda," *Twitter*, October 24, 2016, twitter.com/mikebloomberg/status/790551107936419840.

open ecosystem[38] and has been called the self-driving "national champion" by the country's Minister of Science and Technology.[39] However, these are still early days.

Who is Responsible in the Event of a Problem?

We can analyze this question by referring to the five levels of autonomy defined by SAE and IOMVM. The answer is straightforward for Levels 1 and 2: the driver remains fully responsible. Level 5 is also straightforward: the machine is fully responsible. The answer becomes much more ambiguous at Level 3 (who is responsible in the time between the point when the machine asks the driver to take back control and the moment the driver acts on that request?). At Level 4, the answer is straightforward, in principle, if the circumstances have been clearly defined, and if the onboard computer is able to refuse to take control when it does not consider it to be the right context. The problem is that the context can change along the way: a journey on the highway can start in broad daylight and end at night, rain or fog can become milder or heavy, and so on.

And What About Electric or Connected Vehicles?

The mobility revolution has, in a fairly consistent way, been envisaged as a combination of autonomous vehicles, electric vehicles and connected vehicles.

Autonomous vehicles must be connected to a network in order to receive onboard intelligence updates. But the debate is far from over when it comes to the value of vehicles being connected while in operation. Waymo's vehicles operate exclusively with onboard intelligence, while other manufacturers are expecting a great deal from 5G communication between vehicles and infrastructure. Additionally, some are considering transitional phases during which the vehicle will have no driver but will operate under the supervision of a person located in a remote control center. Finally, if

38 With a few exceptions, such as mapping and machine-learning algorithms.
39 William Keating, "Autonomous driving - Baidu, China's National Champion," Smartkarma, May 8, 2018, www.smartkarma.com/insights/autonomous-driving-baidu-china-s-national-champion.

companies want to implement an in-vehicle movie or advertising system, connectivity would be useful.

Internal combustion engine vehicles can be made autonomous, and several companies are conducting tests with these machines. But the electric engine is easier to control. In addition, since it is autonomous, an electric AV could recharge itself while its owner goes about more interesting tasks. Finally, bearing in mind that revolutions usually don't go half-way we can assume that a vehicle built to modern expectations will avoid leaving a trail of pollution in its wake, and that the issues of CO_2 emissions will be passed on to the power companies.

Where are we Today from a Legislative Perspective?

Until 2015, the almost universal set of regulations governing road traffic was based on the 1968 Vienna Convention, which stipulates, in Article 8, that any moving vehicle or combination of vehicles shall have a driver, and every driver shall at all times have full control of their vehicle. The promise of a significant reduction in accidents through self-driving technology, the strength of global competition between the main players, and the interest of various regions in being used as test areas[40] have led states to loosen the legal framework.

In 2016, the UNECE amended the Vienna Convention to bring it in line with current developments and to allow the signatory states to legislate on self-driving cars. As of now, "automated driving systems" will be explicitly authorized on the roads, "provided that they are compliant with United Nations vehicle regulations and can be overridden or switched off by the driver if necessary." This amendment therefore marks the abandonment of the notion that "the driver shall at all times have control of the vehicle." The official aim of this change is to enable testing, which always involves the presence of a driver who can take back control if necessary.

In France, the Energy Transition for Green Growth Act of 2015 authorized

---

40 Regions' interest in becoming a place for experimentation is linked both to issues of visibility (they will be talked about) and development (see Chapter 6, "Technology").

the government to make changes to legislation in order to allow "partially or fully self-driving cars known as autonomous cars" to operate on the roads. An order of August 2016 authorized testing in these terms: "The operation of partially or fully self-driving vehicles on public roads for the purposes of testing requires a permit in order to ensure the safety of the testing process." With regard to the PACTE law, which was presented to the French National Assembly in October 2018, it proposed a more flexible regulatory framework (that no longer requires an emergency driver to be on board), which aims to make France "the most attractive country in Europe for development of such transformative technology", according to the Minister of the Economy.

In 2016, Belgium drafted a "Code of Good Practices for Testing" which set out the rules and recommendations to be respected when testing autonomous cars in public places. In particular, it stipulates that a person must be in the car to take back control in the event of difficulty during tests on public roads, that vehicles must stay within a speed limit of 30 kilometers per hour (approximately 19 miles per hour) in public places, and that a report must be produced and submitted to the authorities for each test performed.

In 2017, the German Parliament passed a law allowing car manufacturers to test self-driving cars. It establishes a framework for testing, requires the presence of a "supervisory" driver behind the wheel, and makes it mandatory for a black box to be installed in the vehicle. The law will be reviewed every two years to adapt to developments and conclusions resulting from the tests.

Several states in the U.S., as well as South Korea, have taken more flexible approaches to authorizing autonomous vehicle testing. In the United States, the federal government requires private entities to demonstrate the safety of their tests, but only provides an indicative framework for the factors that need to be taken into consideration when conducting this demonstration.[41] In particular, there is no obligation at the federal level to publicly disclose the findings recorded, such as the frequency with which the emergency driver must take back control. California, however, is one state that does impose such a rule.

---

41 U.S. Department of Transportation, Preparing for the Future of Transportation, "Automated Vehicles 3.0," November 3, 2018, www.transportation.gov/sites/dot.gov/files/docs/policy-initiatives/automated-vehicles/320711/preparing-future-transportation-automated-vehicle-30.pdf.

And where do we Stand from the Perspective of Disruption in the Mobility Field?

Consultancy firm Roland Berger[42] has attempted to rank countries according to how favorable their climates are for disruption within the mobility sector, in which the autonomous vehicle plays a central role. The ranking is based on twenty-five criteria, which are divided between five groups: customer interest, regulation, technology, infrastructure, and industrial activity in the sector. The very fact that these five aspects are mentioned tells us that technology alone is not the be-all and end-all for the future of the AV. For example, when we take a closer look, we find key factors in the regulatory dimension alongside the vehicle approval processes, restrictions on the circulation of internal combustion vehicles, and the legislation on carbon emissions. This suggests that innovation will also be achieved by saying goodbye to established practices.

Only ten countries are covered in the study: Germany, China, South Korea, the United States, France, India, Japan, the Netherlands, the United Kingdom, and Singapore. It is regrettable that Finland, which promotes Mobility as a Service (MaaS), is not included in this study, but the results are still very interesting, and surprising too: the two leading countries are those without a domestic automotive industry, the Netherlands and Singapore.

In the Number 1 spot, the Netherlands owes its ranking to its policy of promoting electric vehicles, high sales volumes of vehicles with extensive electronics, strong customer interest, and advanced procedures for the approval of innovative vehicles. Singapore is at Number 2, largely thanks to the emphasis it places on mobility solutions that are not based on car ownership.[43]

At the bottom of the list are the United States and Japan, two countries that were leading producers of automobiles until recently, when they were overtaken by China and its 1.3 billion potential consumers. Their poor

42 Roland Berger, "Automotive Disruption Radar - Issue #1. Tracking disruption signals in the automotive industry," 2017.
43 It should be noted that Singapore limits the size of its vehicle fleet through a quota system, and that anyone who wants to purchase a car must obtain a "certificate of entitlement" at one of the monthly auctions, which is valid for ten years. As a result, it typically costs 70,000 to 80,000 euros to buy a compact car, even before taking usage tolls into account.

performance is due to their weaknesses with regard to electric vehicles and shared vehicles, and the lack of restrictions on the use of internal combustion engine vehicles. As with the Navigant survey, it should be noted that Roland Berger's study is intended to be repeated, and that the current ranking only represents a snapshot of a given moment in time.

Nevertheless, there are still a few reasons to ascribe some weight to this study. One of the two leading countries is a city-state, while the other is a small, densely populated, and highly urbanized country, suggesting that the "mobility revolution" is easier to achieve in dense areas than elsewhere. Another interesting feature is that the Asian countries, where car ownership rates are low, are the ones most heavily invested in vehicle or trip sharing. In these countries, sharing is seen first and foremost as a way of accessing travel by car for people who do not own one, as opposed to a sophisticated solution for those who already do. These are also the countries where individuals who are able to buy a car can most easily consider an electric car, probably because of powerful incentives from governments.

Chapter 2

Cities and Mobility: A Hand-in-Hand Evolution

From getting around on foot to traveling by car, by way of a horse-drawn carriage, a bicycle, and the train, every mode of transport has improved mobility and transformed the urban environment. As travel speeds increased, cities spread and the "mesh" of activities expanded. Options for people were concentrated and aggregated in new ways (e.g., malls and retail centers) over an ever broader geographical area. It is within this sprawling and fragmented city that the autonomous vehicle will have to find its place.

The autonomous vehicle will take its place in a long line of transportation modes that have been introduced successively over the past few centuries. All of them have contributed to the transformation of cities, but not always in the way their designers originally intended. The autonomous vehicle will do the same, as we will discuss later on in this book. For now, we will take recent history as a starting point for our thought process.

Over time, new modes of transport have helped improve mobility and transform urban planning and the mesh of activities across spaces, while extending the city's outer limits, to the point where the very term we use to refer to such places has also had to evolve. At first, we spoke of a "city," then a "conurbation"[44], before progressing to "urban area," and more recently "greater metropolitan area"—and the list is likely to keep growing. The idea that our concept of a "city" is defined by the ability to move around is not intuitive at all. Isn't a city first and foremost a place with high density? Yet this idea of movement refers to the very essence of the city: it is a fundamentally a place for interaction and exchange.

44 An extended urban area, typically consisting of several towns that merge with the suburbs of a central city.

The history of the past two centuries sheds some light on the relationship between the expansion and structure of spaces and the performance of their transportation systems (i.e., the performance of transportation systems in terms of speed, cost, comfort, etc., not that of individual modes of transport). A car is designed to travel at speeds of up to 100 miles per hour, yet the speed of door-to-door travel in cities is less than 25 miles per hour. Despite consistent improvements to the car over the past fifteen years, its speed of travel in city centers has dropped sharply: city-center speeds are a social construct, limited by the robustness of the road network (are there many highways or relatively few?), the number of drivers using the network at the same time (which can be controlled using tolls or charges), traffic and parking rules, and even the status of other means of transportation. How fast would you be able to drive in New York City if the entire MTA transit system broke down, say, or if all the workers were on strike?

In other words, the potential ability of modes of transportation to transform public spaces is linked, on the one hand, to their intrinsic performance and, on the other, to the conditions of the integration of these modes on the roads.[45] The outcome in terms of land use can be seen in a positive or negative light, depending on the individuals and the criteria selected. As far as it can be anticipated, this land use outcome, in turn, becomes part of the evaluation criteria used to determine what place to give a particular mode of transport in the transportation system, particularly when that mode is new and attracting considerable investment, as is the case with the autonomous vehicle.

The Potential for Interaction: the Essence of the City

A city is a collection of various forms or ways of occupying a space that human beings invented to facilitate all kinds of interaction, exchange, and trade. Cities are both places where a large number of people are present in a confined space and places where relations between people are facilitated, which, as a result, makes them more dense and more intense. These are places where ideas, techniques, and know-how are disseminated faster

45 Guided modes of transportation, such as railways or subways, are not included here.

than elsewhere, making them attractive and enabling them to play a driving role in human development.

The essence of a city is contradictory, as perfectly summed up by urban planner Marc Wiel, who observed: "To be urban is to need other people and at the same time to be irritated by them."

When walking was the only or dominant form of travel, cities had to be dense and could only grow by becoming even denser. Lengthening the distances traveled was not feasible, so overcrowding had to be accepted as a necessary downside. But then new means of transportation arrived, and changed everything: "Density is necessary for social and economic interactions when mobility is limited, but it is no longer necessary when mobility becomes easier and less costly... Greater mobility has opened up vast areas to urban spread by offering a range of ways to remain connected to the built-up city center, making it accessible to those who do not live there."[46]

The means of transportation available as history progressed therefore made it possible to build the sprawling cities we know today. The arrival of mechanized transport made it possible to maintain (and even improve) the potential for interaction between urban dwellers, despite the increasing distances that had to be covered, thanks to faster speeds and better transportation capabilities. It is this relationship that we will now explore in more depth, from both a historical and an analytical perspective.

The Joint Production of Space and Mobility: A Look Through History

In the early nineteenth century, cities in Europe and some other countries experienced a boom in population thanks to the first influx of people from rural to urban areas. This change fragmented networks of mutual acquaintance[47], generated a sense of insecurity, and increased pressure

46 Marc Wiel, *La transition urbaine*, Liège, Mardaga, 1999.
47 From the Middle Ages to the French Revolution, people largely stayed in their own neighborhoods: "A gathering place of ancestral practices, each place contains people's daily stampings, encompasses habits of speaking and dressing, and constructs a web of emotional attachments, cordoning off the neighborhoods." And moreover, "Housing is not clearly separated from the traffic area, the street seems to belong more to the inhabitant than to the passerby." Christophe Studeny, *L'invention de la vitesse. France, xviiie-xxe* siècle, Gallimard, Paris, 1995.

on cramped and crowded public spaces. This was a period marked by urban fear, epidemics, social upheaval, and overcrowding.

Initially, in the absence of any collective strategy to counter these problems, members of the upper classes sent women and children to live in safe, healthy, rural areas, or they created new neighborhoods that were expensive enough to ensure a level of homogeneity in the social groups that could afford to live in them. The first short-distance passenger railways then further reinforced this trend, toward the social demarcation of space, by offering the wealthiest urban residents the means to leave the dangerous environment of the city center (the cost of travel was high) and settle in lush, rural areas, away from the urban pressure cooker, while continuing to participate in city life. These same railroads were also used much later to loosen the stranglehold on urban density and to allow people with more modest means to live in ill-equipped and poorly planned suburbs while still working in the city.

Within the cities that had formed at this stage, walking remained the dominant form of travel. However, the need for speed, at a time when increasing numbers of urban residents were starting businesses, meant that vehicle traffic became heavier (including privately owned horse teams, goods carts, horse-drawn buses, etc.). Speed was the future, and pedestrians were to be kept in line by way of sidewalks—and even the coachman's whip!

Thanks to the Industrial Revolution, transportation networks were developed that provided a faster option than travel by horse or on foot. Density was no longer the only factor that facilitated and increased the capacity for interaction, and greater urban density was no longer the only response to population growth. Individuals and governments alike increasingly had the means to choose between the advantages and disadvantages of population density.

Retail also changed. Faced with the irritations of increasing traffic, covered passages, a precursor to our pedestrianized streets, began to proliferate.

With Haussman's renovation of Paris, which served as a model for other cities, public authorities cleaned up city centers, created public parks and gardens, widened existing roads, and built new ones. The increase in road capacity made it possible to accommodate new traffic. Thus, in places frequented by people with greater purchasing power, which were well served by public transportation, department stores sprang up, each with an extensive potential sales area now that trade no longer depended on proximity. One century later, superstores dominated, before they, in turn, were challenged by a renewed interest in local trade and distance shopping.

In the context of class conflict, public authorities decided to move the poorer social classes away from city centers. They provided access to neighboring municipalities via horse-drawn carriage services, and established the practice of "workers' fares" to enable people with limited resources to use public transport to get to their jobs, which were often still in the center of town. At the same time, industrial (often polluting) activities were relocated to these municipalities, where they underwent further growth. Garden cities were built, prefiguring the development of social housing: urban expansion was well underway.

The invention of methods of transportation that ran on electricity or fossil fuels gradually provided faster, less tiring, and less costly types of travel, while the amount of space used per passenger continued to shrink. Thus, streetcars, subways, buses, and commuter trains gradually allowed individuals, whatever their level of living, to reside in a given area and to practice their activities elsewhere. This shift happened thanks to the constant progress made in practicable speeds and to new interconnection capabilities.[48] It was in this context of city expansion and the social demarcation of space that the bicycle arrived, followed closely by the automobile.

The bicycle was first to appear and its use was much more widespread than the car. Both helped solve problems, particularly the issue of the congestion and nuisance caused by horse-drawn vehicles. Until the late 1950s, the number of cyclists and journeys made by bicycle exceeded the number of motorists and car journeys. Yet the car eventually prevailed. Why was that? Initially, the car succeeded because it was the vehicle of the dominant social class, which led to the construction of roads where cyclists soon felt uneasy and at risk. The image of modernity, transmitted by the automobile[49] and its association with progress, led local authorities to develop

48 So, in the suburbs, bus services to train stations greatly expanded the area in which people could look for a plot of land to build a house.

49 In 1951, Renault published a book titled *L'automobile de France*, written by leading intellectuals of the time (Jules Romains, André Siegfried, Jean Cassou, and Georges Friedman). It states: "This unique combination of conditions has meant that now, in the mid-twentieth century, the automobile industry is taking on the nature of an example and an indicator. A country that does not make cars may experience various forms of prosperity and have some advantages. But it cannot be considered a first-rate country." This is the same debate that is being held today around artificial intelligence.

rules and standards that were adapted to the needs of car users.

Decisions that would prove to be crucial to the future of the automotive system and the city were made—without any awareness of their importance. The Decree of 1948 by the Paris Prefect of Police (followed by a 1954 law extending the provision across the country), for example, abolished the obligation to keep position lamps on for vehicles parked on public roads at night, effectively establishing the right to long-term parking on public roads, whereas before (e.g., in the Law of 1851 and the Decree of 1921), the curb was to be used exclusively for dropping off passengers or deliveries.

Then in 1967, France's Land Use Act required new residential and office buildings to build parking lots.[50] Car use became widespread, and private ownership of a vehicle signaled one's membership among the growing lower middle class. Against this favorable backdrop, city residents were quick to "invent the life that goes with it,"[51] using the car to try out new activities and to visit or live in new areas[52], even more so as a large-scale highway program increased the car's ease of use.

In doing this, the issues associated with cycling, in particular those related to the coexistence of cars and bicycles on the roads, were ignored, which made the bicycle a less attractive form of transportation. In the shadow of the automobile, the bicycle missed out on the "snowball effect" (fleets, clubs, and network effects)[53] that made the car so successful. This was borne out by the fact that in the early 1960s, even though there were many more cyclists at the time than motorists, France embarked on the development of a major highway program that was inaccessible to bicycles, and it did this without giving it a moment's thought to creating a bicycle lane program.

The car's rise to dominance not only brought other means of transportation to their knees, thanks to its outright monopoly of the road network, but it also profoundly transformed land use. The cities we have inherited have

---

50 This would have the unintended effect of boosting car ownership and car use for commuting to work.

51 A car ad by Renault for its Twingo model.

52 i.e., while public transportation, which was still very poorly integrated, could take passengers where it already happened to go, the car would take drivers where they wanted to go.

53 Concepts put forward by Gabriel Dupuy, *La Dépendance automobile*, Economica/Anthropos, Paris, 1999. For a recent approach to these effects, see Frédéric Héran, "Le système vélo," *Forum Vies Mobiles*, 2018, fr.forumvies-mobiles.org/reperes/systeme-velo-12437.

been fragmented by the construction of highways and the transformation of existing roads into expressways. The box below summarizes what today's cities owe to motorized transportation.

Ten Results of the Transition from the Pedestrian City to the Motor City

1. More highly populated cities

Major cities owe their entire existence to motorized transportation. The pedestrian city was faced with a problem of demographic limits defined by acceptable density and limited by the maximum walking distance allowed between the city center and housing locations. By enabling the city to expand, motorized transportation virtually eliminated any upper population limit with regard to urban growth. The 12,000 square kilometer (4,633 square mile) area of Île-de-France could accommodate the entire French population with a "reasonable" density of 5,500 inhabitants per square kilometer (14,245 inhabitants per square mile)!

2. Far more sprawling cities that are much less dense

Cities of all sizes expanded thanks to motorized transportation. When urban residents no longer needed to live within walking distance of their places of work, the speed of motorized transportation[54] enabled many more districts and areas that were further away from the center to be integrated into the urban environment, provided that commutes were reasonable (between 30 minutes and 45 minutes). It also allowed residents from the center to move to these remote areas without losing their connection to the city.

3. A sharp decline in the concentration of city-center jobs

The deconcentration away from city centers, which is well known in

54 In terms of numbers, motorized transportation's average speed is about six times faster than walking.

regard to housing, also applies to jobs and businesses to a similar extent, if not more so, even if jobs do tend to move less far afield from city centers. Suffice it to say that Paris now accounts for only 31% of jobs in the Île-de-France region, or that job creation in the United States between 1950 and 1990 amounted to 7 million in city centers and 47 million in suburban areas. Generally speaking, offices tend to cluster together in business districts, while companies providing household services tend to be closer to where people live. In turn, this deconcentration of jobs fuels opportunities for urban expansion, as many residents of the outskirts of major cities no longer have to travel into the city center for work.

4. Cities have become a patchwork of differentiated areas

The historical view of the city as a monocentric, homogenous space (excluding natural barriers, uniformly decreasing in density from the city center to the outskirts, and with equivalent possible speeds of travel in all directions) is no longer applicable today. Certainly, in Europe, the multi-functional city center still exists, but it is increasingly a space for tourism and leisure, and decentralized employment has become the norm in large cities. Major urban areas have enveloped smaller towns, creating local pockets of population density, absorbing suburbs, developing new residential areas, and sometimes creating new cities. The speed of travel that can be achieved at a given place in a region is highly dependent on the presence (or lack) of high-speed infrastructure (e.g., urban highways, regional railways). In addition, the accessibility of a given area relies heavily on the presence of a station or transfer hub. Lastly, the effect of the infrastructure on the areas being traveled through extends far beyond their control on the ground (e.g., areas where housing construction is prohibited due to noise pollution).

5. Cities whose development became partly independent from their roadway fabric

Historically, cities developed exclusively along a grid of roads, and by forming individual parcels of developed land of a relatively uniform size. Since the 1960s, we have seen the development of vast areas (i.e., social housing, racket-shaped suburban developments, vast commercial business zones, and campuses) that have been freed from this constraint and are accessible only through entrances/exits that are generally designed for cars.[55] These areas, along with high-speed infrastructure and non-traversable roads, contribute to the fragmentation of the urban fabric.

6. Increased social segregation

The city of the past was a place where people's commutes to work were short, due to limited mobility. As a result, people of different social statuses who were employed in the same workplace (e.g., a foreman and a skilled worker) lived near each other. This need for proximity to one's employment has now disappeared. Increasingly, the wealthiest social classes are able to choose their place of residence regardless of their place of work, and instead make their choice based on the quality of local schools, services, access to cultural spaces, and even certain landscape features. In doing so, they tend to group together in upscale neighborhoods. This drives up real estate prices (as the value of a real estate property depends on the income of neighboring households) and contributes to driving away people of lesser means.

7. Massive improvement in average living conditions

It is probably in this area that progress has been the most significant. The majority of post-war housing lacked many modern conveniences (such as bathrooms) that have since become the norm. In terms of size, today's housing continues to grow in France (980 square feet per housing unit compared to 883 square feet thirty years ago), while the average number of occupants per household continues to decline (2.7 compared to 2.3). These developments are all the more remarkable (even if the French average remains below the American norm) since they occurred in a context of migration from the countryside (where

---

55 Here, we see the opposing principles so clearly illustrated by David Mangin of the "walking city" and the "franchised city" where all of the land (roads included) is privately owned. See David Mangin, *La ville franchisée*, Éditions de la Villette, Paris, 2004; David Mangin (ed. by Ariella Masboungi), *La ville passante*, Éditions Parenthèses, Marseille, 2008.

space is inexpensive) to cities (where it is very expensive). The trends in housing size are similar in the U.S. (2,687 square feet in 2015 compared to 1,700 square feet thirty years ago), while the average number of occupants per household also continues to decline (2.69 in 1985 compared to 2.54 in 2015).[56]

These developments could only occur with sharp increases in household spending on housing, offset by an urban sprawl that has made it possible to curb housing costs. However, this overall trend of improvement does not preclude the problematic issues of poor quality housing and overcrowding, usually linked to disadvantaged and poverty-stricken households.

8. Access to jobs and urban amenities has evolved in different ways

In a region as vast as Île-de-France (the region that includes Paris), a study carried out in the early 1990s showed that an active person could access three-quarters of the area's jobs within less than an hour, on average, via the fastest mode of transport (car or public transit)—a remarkable finding for an area of 12,000 square kilometers (4,633 square miles).

It is unlikely that this has ever been achieved in the past, even when the city was more compact, and it is unlikely to be the case in the future, due to traffic reduction policies (and specifically lower speed limits for vehicles in city centers). Performing the same analysis on everyday services, particularly retail shops, is somewhat more difficult. Grocery stores and superstores have contributed to a significant reduction in local trade. While this may not be a problem for urban residents who drive, it is for those who do not own a car or are no longer able to drive (e.g., the older population) and whose access to stores may be greatly reduced.

56 The American Enterprise Institute, a public policy think tank dedicated to defending human dignity, expanding human potential, and building a freer and safer world. Blog Post by Mark J. Perry, Scholar, June 5, 2016, https://www.aei.org/carpe-diem/new-us-homes-today-are-1000-square-feet-larger-than-in-1973-and-living-space-per-person-has-nearly-doubled/

9. Gridlocked cities

Ease of movement and transportation is at the root of the spread of congestion from city centers to wide swaths of a metropolitan area. This is a misunderstood paradox: congestion is due to the new organization of urban areas, which was produced by a reduction in the costs of all of the components of mobility. Increasing the capacity of a road or a congested urban area generally only solves a congestion problem temporarily. As an induced circulation increases, the new roads will gradually end up being as congested as the ones they were originally intended to relieve. This is what Martin Mogridge described as the "self-defeating" nature of urban road capacity policies[57]: they are at the origin of global traffic growth and generally fail to keep their promise of reducing congestion.

10. Polluted cities

Road traffic, which continues to increase and relies on oil-derived products as its main source of fuel, contributes to greenhouse gas emissions and is responsible for a significant and growing share of urban air pollution worldwide.[58] Other nuisances and kinds of pollution—sound and visual—are added to the former. That being said, ease of transportation is not exclusively damaging to health: by allowing better access to jobs, it can also have positive effects...which in turn can generate secondary negative effects. For example, the car has taken market share away from more active modes of transport (e.g., cycling, walking) that are beneficial to health.

In sum, the city of today, which was built by and for mechanized mobility, has seen the goals of the health-conscious, nineteenth-century urban planners come to fruition. It has shifted places of residence away from places of employment (which at the time were noisy and polluting factories), moved workers who may be tempted to

---

57 Martin J. H. Mogridge, "The self-defeating nature of urban road capacity policy: A review of theories, disputes and available evidence," *Transport Policy*, Vol. 4, No. 1, January 1997, p. 5-23.

58 Deutsche Gesellschaft für Internationale Zusammenarbeit (GIZ) on behalf of the German Federal Ministry for Economic Cooperation and Development (BMZ), Urban Transport and Health, 2011; Urban Transport and Climate Change, SUTP, 2014.

revolt away from iconic sites of revolution (the city center), and offered these workers more suitable housing conditions, thus decreasing the density of cities to reduce the transmission of infectious diseases. The paradox here is that their project goals really reached their conclusion at a time when factories have disappeared from cities—the appetite for revolt has hardly ever been lower, and infectious diseases have become much more rare.

The Joint Production of Space and Mobility: The Mechanisms at Work

By observing mobility in highly varied contexts (e.g., in small and large cities, as well as developed, emerging, and developing countries), Yacov Zahavi demonstrated while working as a researcher at the World Bank in the 1970s that the amount of time and money people spend on transportation is a type of constant. What we now know as Zahavi's conjecture or law[59] establishes that generally all over the world, and at all times when mobility could be measured, people spend an average of just over one hour a day traveling. His research has since been confirmed by several international studies, one of the most complete of which was carried out by A. Schäfer and D.G. Victor on a large sample of cities, villages, and countries.[60] Its conclusion was that the average travel-time budget is around 1.1 hours per person per day, with a standard deviation of 20 minutes.

As a result, any increase in the speed of available transportation leads to an increase in the distance traveled in the medium term, with a constant travel time budget (TTB). The time saved thanks to a higher speed is reinvested with the goal of accessing new opportunities, such as living in larger housing that would previously have been too far away (in terms of travel time) from a person's workplace.

---

59 Yacov Zahavi, *Urban Travel Patterns*, The World Bank, 1980.

60 Andreas Schäfer and David G. Victor, "The Future Mobility of the World Population," *Transportation Research*, Part A 34, 2000, pp. 171-205.

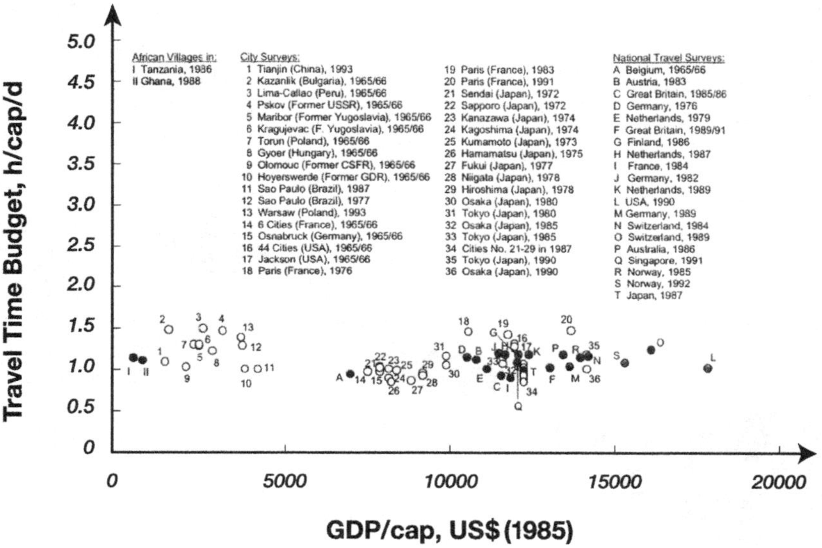

Figure 3. Daily Travel Time in Hours per Person based on GDP per capita

Source: Andreas Schäfer, David G. Victor, op. cit., p. 175

This reading of our behavior gives us a key to understanding the dynamics of the somewhat "diabolical" relationship between mobility and the city.[61] The increase in travel speeds that came with the emergence of the railways in the nineteenth century, followed by the automobile and the emergence of highway networks in the twentieth century, and combined with a wealth of available pieces of land on the outskirts of cities, produced changes in the morphology of our cities (which spread), the localization strategies of its stakeholders (such as inhabitants, developers, traders, and government administrations), and their ways of life or functioning.

In other words, the offer of modern transportation created demand; our demand—without us being aware of what was happening—and thus our mobility ended up moving the very places themselves, from local neighborhood shops and movie theaters to shopping malls and giant multiplexes.

61 Title of the book by Marc Wiel, *Ville et mobilité: un couple infernal?* ["The City and Mobility: A Diabolical Couple?"] Éditions de l'Aube, Paris, 2004.

Chapter 3

The Autonomous Vehicle: A Timely "Disruption"?

Over the last two centuries, each generation has seen a new mode of transportation arrive and become widely adopted. Yet, in the past fifty years, there were no major innovations to report—until the arrival of the autonomous vehicle was announced.

While the term "disruption" has become part of our modern lexicon, as a way of referring to a period of upheaval, it must be noted that in terms of mobility, nothing as disruptive as rail, car, or air travel has appeared in a long time. Meanwhile, the systems currently in place (i.e., public transportation and car use in cities) are showing signs of running out of steam and have even become sources of dissatisfaction. What if the next revolution in mobility were the autonomous vehicle?

A Lack of Innovation and Disruption in Transportation at the Start of the New Millennium

For the past two hundred years, each generation has seen the arrival of a new mode of transportation, or at least has seen one spread to every aspect of society. In cities, the nineteenth century witnessed the progression from horse-drawn carriages to streetcars and then to subways, as well as the broader transition from stagecoaches to railways.[62] It also saw the emergence of individual mobility in the form of the bicycle.

62 It is this progression that Schumpeter gives as an example of what he calls "creative destruction," a concept that is close to what we now know as disruption: "Add successively as many mail coaches as you please, you will never get a railway thereby." J. Schumpeter, 1934, The Theory of Economic Development, Harvard University Press, 1949.

The twentieth century saw the arrival and adoption of buses, motorcycles, cars, and planes as well as their improvement over time (e.g., the appearance of highways, high-speed trains, and low-cost airlines). Upon reaching adulthood, an individual could feel that they were traveling in a better way than their parents or grandparents had, and that was no illusion. Research by the IIASA[63] shows that each major technological revolution has also brought about strong growth in mobility by creating new opportunities for travel as well as faster means of doing so. Since 1880, the average annual distance traveled by a person in the United States has risen by 2.7% per year. In France, the same figure has increased by about 4% since 1800 (not including travel by foot).

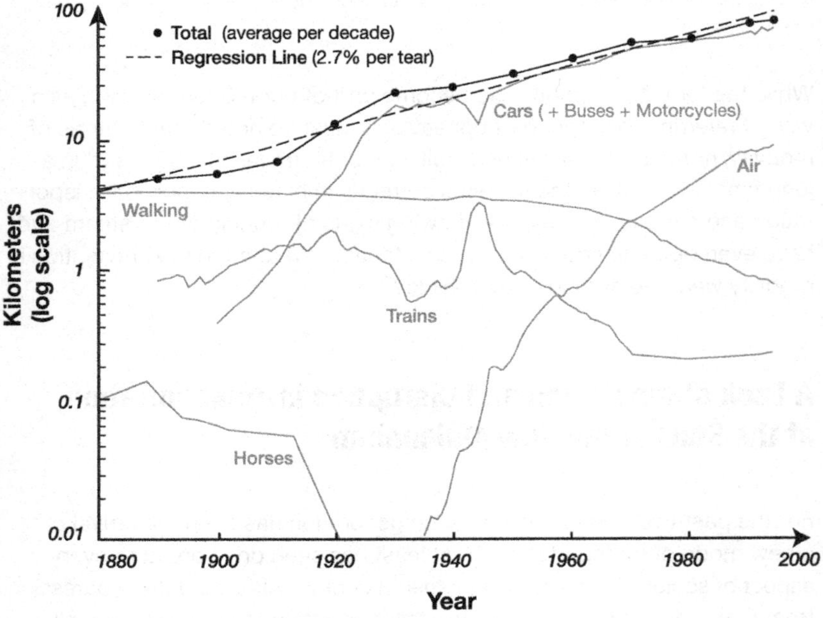

Figure 4. Change in distances traveled per person in the United States since 1880

Source: Jesse H. Ausubel, Cesare Marchetti, and Perrine Mayer, "Toward green mobility: The evolution of transport," European Review, Vol. 6, No. 2, 1998, p. 142.

63 International Institute for Applied Systems Analysis

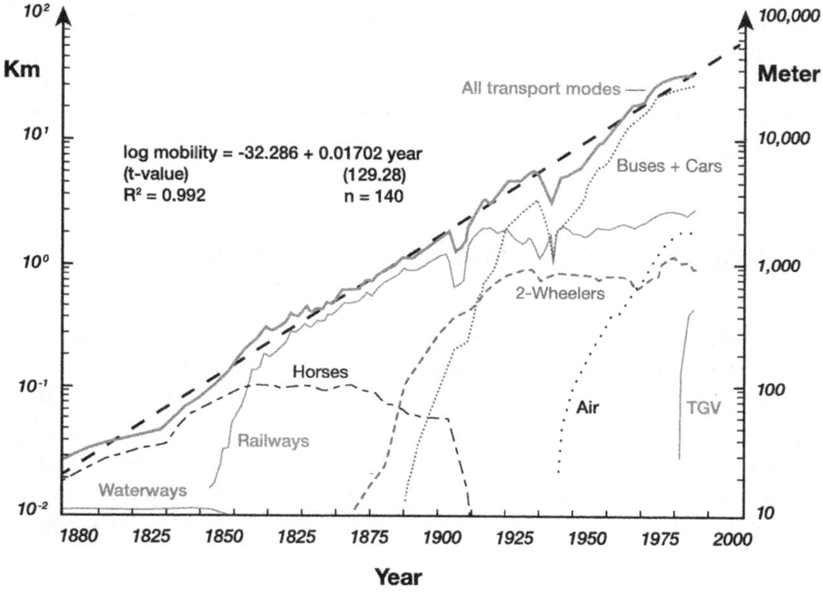

Figure 5. Change in distances traveled per person (not including travel by foot) in France since 1800

Source: Arnulf Gruebler, The Rise and Fall of Infrastructures: Dynamics of Evolution and Technological Change in Transport, Heidelberg: Physica-Verlag, 1990, p. 232.

Major innovations (e.g., railways, automobiles, commercial airlines) have occurred approximately every fifty years. A half-century is also the minimum amount of time it has taken for them to reach their full potential.[64] Looking back over the last fifty years (from the late 1960s to the present), it is difficult to identify a major innovation in transportation. The high-speed train may well represent a "disruption," but it has only made it possible to develop markets that had already been opened up by air travel (and highways, to a certain extent). The streetcar has had a welcome return, and it has helped ease traffic in urban areas, but it is just a drop in the ocean of mobility. Yes: cars have been vastly improved, and we can take phone calls in our cars and have more choice in music, but at no point have they truly surprised us or led us to

64 By way of example, it was not until the late 1960s that the automobile, which had emerged at the turn of the twentieth century, finally became a fixture in half of French households. (Although it is true that both World Wars delayed its adoption.) For air travel, it took the arrival of jets, jumbo jets, and chartered planes for this mode to no longer be seen as belonging exclusively to the elite.

imagine new possibilities. A Renault car ad may urge us to "invent the life that goes with it," but the car of today hardly broadens people's horizons any more than its 1960s counterpart.[65] The facts speak for themselves: a person on the cusp of adulthood today only has access to the same means of transportation as their parents (or even grandparents).

Established Systems have Reached an Impasse

The IIASA has interpreted the strong growth in mobility seen in the past, measured in terms of distances traveled, as an indication of the expansion of our "ecological niche." Yacov Zahavi has added to this approach by proposing the travel time model of human behavior mentioned in the previous chapter[66]—i.e., we seek to make the most of the resources that are available to us (be they jobs, businesses, or public or private services), provided that the mobility solutions remain within a budget that is acceptable in terms of cost and travel time.

The growth in mobility has also been accompanied by a significant shift in public infrastructure policies. Until the 1980s, public authorities attempted to keep up with demand: since people were buying cars, they had to be provided with suitable roads they could use. This approach has been increasingly challenged, especially in Europe, for political reasons, particularly objections to the construction of expressways in cities, and a wider-scale opposition to high-speed infrastructure (including railways), due to environmental concerns (such as pollution, noise, and contribution to climate change).

Such challenges were considered acceptable by the transportation engineering community, which has given up on raising the specter of crippling congestion, thanks in large part to Zahavi's findings: if the infrastructure necessary to maintain or increase speeds is not built, then demand will stagnate, drop, or shift to other modes (e.g., public transportation, cycling) that require less space. This is what Susan Owens has called the transition from a "predict and provide" strategy to one of "predict and prevent."[67] This transition entailed the exit of mobility from the usual economic refer-

---

65 However, it can be used in other ways, with rental and self-service systems, and we will come back to this later.
66 Yacov Zahavi, *The UMOT Model*, Department of Transportation, Washington, D.C., 1980.
67 Susan Owens, "From 'predict and provide' to 'predict and prevent'?" Transport Policy, Vol. 2, No. 1, 1995.

ence, which is limitless or unlimited (for example, the GDP: always more!), and its entry into the world of limits. This shift is also what others have called the disappearance (or evaporation) of traffic: that when a major road is closed, only some of the traffic it supported is transferred to other roads; the rest disappears. This entails the (somewhat questionable) belief that flows will adapt to the size of the pipe.

This new approach may make sense when it comes to fluid mechanics, but from a sociological perspective it lacks depth. The increase in distances traveled also includes a number of hidden factors: family relationships or friendships being maintained despite a move to a big city, perhaps a summer house used for weekends and holidays, accepting a more attractive job because it was relatively easy to get to, regular trips to a favorite nature reserve with the children, and so on. Ultimately, one or more of these activities has to be given up when the size of the "pipe" is reduced, and that is frustrating. This trade-off serves as a reminder that there are always two aspects to mobility: the mobility that is achieved, as reflected in the traffic actually observed, and a latent demand for mobility, which will be expressed if, for example, the mobility conditions offered by a new system make the new demand possible.

From a traffic engineering standpoint, this new approach, which is barely a quarter of a century old, has produced indisputable and expected results. Passenger car traffic has decreased in many city centers. Use of public transport has increased, but at the cost of an increase in operating subsidies, especially in France, which is financially difficult to sustain for local authorities. Moreover, the increase in usage sometimes leads to overcrowding on transit vehicles, which can create a feeling of going from bad to worse.

This new approach has also produced less desirable, or less acceptable, effects, none of which have attracted much media attention. For example, in the Île-de-France region (greater Paris), the latest survey on mobility has shown that for the first time since surveys have been conducted, the average speed of travel has fallen, for both cars and public modes of transport.[68] This decline is all the more unwelcome, given that calls for mobility are as loud as ever, that the dynamics for turning cities into metropolises

68 Jean-Pierre Orfeuil, "Les déplacements en Île-de-France: vitesse au vingtième siècle, lenteur au vingt et unième?" Revue Transports, No. 502, 2017.

have tended to integrate urban functions in ever wider spaces, and that the rise in real estate prices continues to motivate segments of the population to move away from city centers.

The policies implemented in the United States have been remarkably different so far: the role of the car has not been affected, or if so only marginally. After a slight decline following the 2008 crisis, and then a six-year period of stagnation, road traffic has picked up again and is now breaking new records, as are levels of congestion. In contrast, public transportation use is facing the opposite problem, having been in continuous decline since 2014.[69]

69 American Public Transportation Association, *Understanding Recent Ridership Changes*, 2018.

Always Going Further and Faster: Is that all Over Now?

From the nineteenth century on, mobility (in terms of distances traveled) increased in line with improved standards of living, without any rise in travel times, due to gains in speed and capacity that were made possible by revolutions in transport systems. The situation seems to have changed since the new millennium, specifically in developed countries where the total number of cars on the road is close to the saturation point, and where investment in road infrastructure is declining. Does this mean that developed societies have slowed down? The Internet and air travel—not included in this study—would suggest otherwise, but a paradox remains: large metropolitan cities increasingly concentrate the global economic value creation thanks to speed and acceleration, but we travel within these cities at slower speeds than elsewhere, and more and more slowly.

	1969	*1990*	*2001*	*2009*	*2017*[70]	*AAGR* 1969–2001*	*AAGR 2001–2017*
Number of trips/person/year	737	1,372	1,493	1,383	1,230	+2.2 %	-1.2 %
Distance covered/person/year (miles)	7,121	12,742	13,465	13,187	14,228	+1.9 %	+0.3 %
Number of car trips/driver/year	847	1,190	1,223	1,102	986	+1.2 %	-1.3 %
Distance covered by a car/driver/year (miles)	7,534	10,399	11,946	10,574	10,399	+1.5 %	-0.9 %

Table 1. Changes in Mobility in the United States

**AAGR: Average annual growth rate. Note: These surveys do not reflect long-distance travel involving nights away from home. Sources: NHTS (National Household Travel Survey), Summary of Travel Trends, Department of Transport, 2009, for data up to 2009, and nhts.ornl.gov for 2017.*

	1973	*1990*	*2000*	*2010*	*2015*	*AAGR* 1973–2000*	*AAGR 2000–2015*
Number of trips/person/year	956	1,094	1,074	961	914	+0.4 %	-1.1 %
Distance covered/person/year (miles)	4,476	6,506	7,159	6,689	6,649	+1.8 %	-0.5 %
Duration of trips (hours/year)	353	365	380	370	368	+0.3 %	-0.2 %
Speed of travel (miles/hour)	12.7	17.8	18.8	18.1	18.1	+1.5 %	-0.3 %

Table 2. Changes in Mobility in Great Britain

**AAGR: Average annual growth rate. Note: These surveys do not reflect long-distance travel involving nights away from home. Source: National Travel Survey, various dates.*

70 The 2017 data is adjusted by the Federal Highway Administration to make them comparable to previous years' results, despite changes in the method used to calculate the distances traveled. See U.S. Department of Transportation, Federal Highway Administration, National Household Travel Survey, Summary of Travel Trends, 2017.

	1976	*1991*	*2001*	*2010*	*AAGR* 1976–2001*	*AAGR 2001–2010*
Distance/person/day excluding walking (km)	12.29	15.51	15.91	16.55	+1.0 %	+0.4 %
Duration/person/day excluding walking (hours)	58.3	66.0	67.4	73.7	+0.6 %	+1.0 %
Speed (km/hour)	13.4	14.1	14.8	14.0	+0.4 %	-0.6 %

Table 3. Changes in Weekday Mobility in the Île-de-France Region (includes Paris)

**AAGR: Average annual growth rate. Note: Distances are "as the crow flies." Sources: Résultats Détaillés, Enquête Globale Transport 2010 (Detailed Results, 2010 Global Transport Survey), OMNIL (Observatoire de la mobilité en Île-de-France). Only covers journeys within Île-de-France.*

The situation in non-metropolitan regions is quite different but no less worrying. Resources to reduce car use are more scarce, and traffic is less of a problem. However, residents of such areas do face other, less visible, less publicized problems, which can lead to deep-seated resentment.[71] These individuals are very dependent on using cars. They are obliged to accept the cost increases, sometimes justified by urban issues alone (certain pollution standards, for example), without being able to switch to other transport modes. They must listen to anti-automobile speeches of clever and trendy urbanites. In the areas where populations are declining and jobs and services are becoming scarce, making trips that are always getting longer—and therefore more costly—is required simply to live one's life. Finally, a widening gap is forming when it comes to the public burden for funding transportation: non-urban residents are paying full price for their travel, while urban residents, who use public transportation, pay at most a quarter of the real costs of the services they use. This is just one aspect of the regional divide.

71 These became loud and clearly visible with the "gilets jaunes" or "yellow vest" protests in France in November 2018.

In Search of a Feasible Emergency Exit

This may seem a particularly bleak picture, but based on the current economic environment, it will probably continue to worsen. There are several reasons for this trend.

- Regional dynamics, driven by the knowledge economy, tend to result in the concentration of development in metropolitan areas[72], the abandonment of non-metropolitan regions, and therefore the widening of the regional divide, all within the context of global competition. This leaves public authorities with little room for maneuver to "rebalance the country," as they have done in France, for example, with some success in the past.
- Metropolitan areas are no longer structured (or no longer purely structured) with the aim of facilitating connections with suburbs that can be fairly easily served by mass transportation. Peri-urban spaces are emerging, which may even play a prominent role in some countries (such as "edge cities"73 in the United States, for example). Traditional public transportation is struggling to meet the more diffuse demands of these regions.
- The requirements of the ecological transition are likely to only become more pressing, so the pressure to reduce solo driving—the most practical mode in at least 90% of cases—should also increase.

In short, if nothing changes, we will be forced to become increasingly agile and flexible, while working with an increasingly creaking or costly system. Not the most optimistic outlook, it must be admitted.

And yet, the future doesn't have to be bleak, for three reasons. First, the difficulty we are facing is less a problem of the city than a manifestation of its very essence. The history of the city has always been one of "trade-offs between the disadvantages of crowding and those of longer journeys,"[74] "of the relationship between the individual and the collective,

72 As reflected, among other things, in the continued rise in property prices in these areas around the world, which is hindering their long-term development.
73 Joel Garreau, *Edge City: Life on the New Frontier*, Doubleday, Anchor Books, 1991.
74 Marc Wiel, *op. cit.*, 1999.

of an acceptable or hoped-for balance between the 'me' and the 'us.'"[75] Second, the range of documented solutions, some of them implemented here and there with a certain degree of success, is constantly expanding. Never before have we had so many ways to solve our problems.[76]

Third and finally, many innovative solutions are being invented by the younger digital generations (see box below). Some of them rely on "their" device (the smartphone) to tinker with and combine various existing means of transportation (e.g., car sharing, carpooling, self-service vehicles, and on-demand transportation). Others are relying on their culture—the culture of globalization—to introduce micro-mobility options, from the hoverboard to the scooter.

75 Marc Wiel, *Pour planifier les villes autrement*, L'Harmattan, Paris, 2007.
76 We only need think of the modal solutions that did not exist in 2000: electric scooters, self-service bicycles, electric bicycles, car sharing, carpooling, ridehailing platforms, and peer-to-peer car rental platforms. Or similarly, we can consider the congestion charge, which was introduced in London, and was considered and subsequently postponed for New York, but never mentioned in Paris, as well as the carpool lanes that have existed on U.S. outer urban highways since the 1970s, and many other types of regulations around the world.

What will the Citizens and Consumers of Tomorrow be like?

"All of modernity is speed, acceleration, productivity gains. The drive for consumption will not simply disappear. Novelty is not an invention of the advertisers. It is driven by the deep inner workings of modern democratic society. Tocqueville understood all this one hundred and fifty years ago. Rather than turning to the heavens or the past, people are looking to the future for answers, and the perpetual creation of new things is part of that."[77]

One hypothesis put forward suggests that an anthropological transition

77 Gilles Lipovetsky, interview in *Le Figaro*, "L'heure est aux rêves de moins et mieux," April 24, 2018; about his new book: *Plaire et toucher, essai sur la société de séduction*, Gallimard, Paris, 2018.

is leading younger generations that grew up in the digital era ("digital natives," also dubbed "millennials") to make different consumer and travel choices than the generations before them, who were influenced by Fordist mass production and the preference for owning objects that signified status (including the automobile). This transition has lent weight, in the United States,[78] to the idea of a quiet transportation revolution, driven by the arrival of new generations that are fully harnessing the power of apps for the purposes of shared mobility. These studies have often made the mistake of taking an overly individualistic approach (and therefore ignoring the relative impoverishment of young people compared to other groups, which makes home and car ownership a luxury for some) and have at times succumbed to wishful thinking.[79] However, the fact remains that worldviews, values, and aspirations are changing, and they do have an impact on people's lifestyles and mobility.

If we want to determine what kind of society the autonomous vehicle will be introduced into, we must bear in mind that the individuals making up this society will be highly diverse, and increasingly so over time (due to growing economic inequality, the widening age gap, the metropolization process, the revitalization of rural life, and differing perspectives on the world). However, they will share new ways of living, which we can attempt to grasp by exploring individual archetypes[80] that are adapted to the world of the future. At the risk of caricature, we will outline below the character traits we expect to see increasingly in the citizen of the future.

The citizen is freed from collective constraints and is somewhat unpredictable. "The dynamics of commoditization and hyper-individualization generate increasingly disorganized, fickle and nomadic behavior."[81] Better educated and informed than their predecessors, the citizen of the future is permanently connected, prefers peer-based information

78 See, in particular, among dozens of other references: APTA, *Millennials and Mobility, Understanding the Millennial Mindset*, 2013. Also Frontier Group, *Millennials in Motion: Changing Travel Habits of Young Americans and the Implications for Public Policy*, 2014.

79 See Jean-Pierre Orfeuil, "Les *millennials* portent-ils une nouvelle approche de la mobilité?", Think Tank URF, 2015. Also, Jean-Pierre Orfeiul, La Fabrique de la cité, "*Millennials*, une légende urbaine," 2017, www.lafabriquedelacite.com/projets/les-millennials-une-legende-urbaine.

80 See, for example, François Ascher, *La société hypermoderne*, Éditions de l'Aube, Paris, 2005.

81 Gilles Lipovetsky, *op. cit.*

(feedback from friends, acquaintances, and the community) over top-down information (from political, religious authorities, or experts), and prefers weak ties to strong ones.[82] In a faster-paced world, (slow) reasoning faces competition from (fast) emotion[83]—the individual gives priority to feelings; he or she is open to narrative knowledge, valuing the ideals of justice, happiness, and beauty on the same level as trust.[84] Innovations may become widespread through the use of effective narratives rather than through demonstration.[85] The citizen of the future is not unsociable, but he or she belongs to groups that can be left at any time. His or her trajectory is less defined by a predetermined course. It is often revised, while flexibility and the ability to bounce back are essential qualities. Shrinking time frames are accompanied by an expansion of horizons with regard to space: it is more common to study or work in locations that were previously seen as the other side of the world, and to compare the merits of transportation policies of London, Tokyo, New York, or Paris. Flexibility requires a certain "lightness"[86]: like the individual's body, which is preferably slim, this is also reflected in the items they use (e.g., smartphones, digital music files) and their relationships with those things (the age of access as opposed to ownership), relationships to others (a certain coolness), and relationships to politics (flexible advocacy rather than being a card-carrying member of a party). Health, the environment, and quality of life all take on increasing importance, in contrast to the ownership of material goods.

In the liquid society announced by the sociologist Zygmunt Bauman[87], the individual is a bit like Jules Verne's Captain Nemo, "Mobilis in mobile," a trajectorial individual.[88] However, the individual still needs points of reference, as well as systems and frameworks that provide reassurance and a sense of trust. In the consumer goods sector,

---

82 On this concept, see Mark Granovetter, "The Strength of Weak Ties," *American Journal of Sociology*, Vol. 78, No. 6, 1973, pp. 1360-1380.
83 Daniel Kahneman, *Thinking, Fast and Slow*, Farrar, Straus and Giroux, New York, 2011.
84 See Jean-François Lyotard, *La condition postmoderne*, Éditions de Minuit, Paris, 1979.
85 Christian Salmon, *Storytelling*, La Découverte, Paris, 2007.
86 Gilles Lipovetsky, *De la légèreté*, Grasset, Paris, 2015.
87 Zygmunt Bauman, *Liquid Modernity*, Polity Press, Cambridge, 2000.
88 This term was coined by Alain Ehrenberg, *L'individu incertain*, Calmann Levy, Paris, 1995.

brand names have traditionally played this role, but they pay a high price when they betray our trust (the Volkswagen scandal, for example). In the services industry, platforms such as Airbnb, Uber, and BlaBlaCar have popularized the practice of peer rating. As far as the regions around cities are concerned, the preference for gated communities by the upper social classes, the avoidance of disadvantaged areas by the middle classes[89], and immediate protests against anything that threatens to disrupt the existing order (e.g., the NIMBY and LULU[90] phenomena) provide the necessary reassurance.

It is difficult to ignore a correlation between, on the one hand, the perpetual search for novelty, flexible behaviors, and the search for "lightness," and, on the other hand, the revival of the bicycle and the emergence of platforms that offer new ways of getting around. However, it is worth keeping in mind that this changeable individual will only remain loyal to a type of transportation if it provides a suitable travel experience.

89 On this point, see Jacques Donzelot, "La ville à trois vitesses," *Esprit*, No. 303, March-April 2004.
90 NIMBY: Not In My Back Yard; LULU: Locally Unwanted Land Use; NOOS: Not On Our Street; CAVE: Citizens Against Virtually Everything.

Many of these solutions combine both functional and playful aspects and meet expectations with regard to health and wellbeing. The first group of solutions have an interest in finding a way out of the conflict between instrumental rationality ("the car is practical") and axiological, value-based rationality ("the car is evil"), which has a negative impact on car usage, but they have more detractors than admirers. The second group of solutions may be viewed with a certain degree of condescension (e.g., the bicycle until recently, the scooter today). None of them have received the stamp of approval of a major public authority or key political leader, who continue to interest themselves more in infrastructure than in systems.[91]

91 However, it should be noted that the LOM Law (Loi d'orientation des mobilités, the French Framework Law on Mobility), prepared by the mobility forum "Assises de la mobilité," mentions these modes and attempts to find a place for them in the mobility landscape.

The problem is not that there are not enough solutions; it is that there are too many. The challenge is first and foremost a cultural one. We must accept that we will not move from one quasi-monopoly (the private internal combustion engine vehicle) to another one, whatever it may be, and that the future lies in a myriad of solutions. We are beginning to understand this in the energy sector, so there is no reason we should not do the same in the mobility field. We will need to find players that are willing to support each of the solutions, but without claiming a hegemony. Then these players will need to coordinate with very different cultures[92], through a holistic perspective that focuses on providing an alternative system to solo-driving. Finally, we will need to estimate the overall role of these new services in the future mobility landscape of each region, without trying to respond to all use cases, which would clearly not be possible.

The problem is to somehow find a way to disassemble the system of self-locking paving stones: the infrastructure organized around the personal car/public transport duo. Major crises can do just this—for example, the New Deal in the United States following the 1929 crisis, or the welfare state in France during the Resistance. In the urban transportation sector, projects[93] and objects (e.g., in the recent past, streetcars), or what Bruno Latour calls "objects with agency,"[94] which have a major impact on society and thus have the ability to change perceptions, beliefs, and roles and can reconfigure the relationships between the players involved.

The Autonomous Vehicle: The Catalyst for Redesigning our Transportation Systems?

An "object with agency" may also be a kind of UFO, which has an impact even without being implemented by a political authority. Social networks and smartphones have accustomed us to this.[95] The autonomous vehicle

92 This will range from the green pioneers of car sharing to the cowboys of the ridehailing world, via the captains of industry such as Bolloré with Autolib, the start-up inventors of digital hitchhiking, and ordinary citizens who enjoy the autonomy offered by the electric bicycle.

93 See Frédérique Hernandez, *Le processus de planification des déplacements urbains entre projets techniques et modèles de ville*, Phd Thesis, Université d'Aix Marseille 3, 2003.

94 Objects which are not mere resources or constraints between people, but which have the ability to act and interact with people.

95 But let's not forget all the innovations that haven't found any takers; there are even entire museums in their honor!

could follow the same model: it has no serious competitors in the field of radical urban mobility innovations; it has existed in our collective imagination for a long time; it can be (and already is) creating a seductive narrative. This narrative appeals both to our instrumental rationality (it is practical) and to our axiological rationality (it will be good for the city, the environment, and the planet). It is aligned with a key aspect of today's "no worries" society: "I call it, it comes and it takes me where I need to go, and that's it!"[96] The potential synergies with other existing innovations (e.g., connected vehicles, electric vehicles, car and ride sharing solutions, and navigation systems), or in the early stages, with vehicles providing mobile activities, are considerable.[97] There are downsides, of course, which we will look at in detail later. Let us simply note here that with this vehicle, our lives (and not just those of their users) are being entrusted to artificial intelligence, which is fallible and "hackable"; our personal data may travel more than we do; and in an autonomous vehicle, it is we who lose our driving autonomy.

However, there is a significant difference between self-driving cars and the ways in which the objects and systems above are used. Using the Internet or a smartphone entails a private relationship between a customer and a service provider. At first glance, such usage has no impact on others.[98] It occurs in a space—a virtual space—that has been created entirely by tech innovators. By contrast, the self-driving car will only become part of our day-to-day lives by occupying a place in physical space. Highways and expressways aside, this space is also seen as a public space, a place for mingling and impromptu encounters, a place that furthers civic spirit, where society meets and convinces itself that we can co-exist. It is not that the state has seized on the self-driving car while the market wasn't looking; it is something to be used and enjoyed by all—a common good that is naturally regulated by public authorities.

96 Jean-Pierre Le Goff, interview in *L'Express*, February 2018.
97 On this subject, see the Mobile Hyperplaces project of the City on the Move VEDECOM Institute project, www.mobilehyperplaces.com
98 This is patently false, particularly due to the network effect, the disappearance of phone booths being just one example. Taking the observation a step further, the fact that this shared object has been replaced by a personal device, the cell phone, shows us that the shared economy is not the only direction we are headed in.

Chapter 4

The Autonomous Vehicle: Huge Benefits Expected but Not Without Controversy

Could the autonomous vehicle be the miracle pill to cure urban traffic congestion and global warming? Here is an in-depth summary of the various ideas on what the self-driving car could offer us.

Although the autonomous vehicle is not yet a fixture in our daily lives, it has already generated plenty of commentary. There are millions of web pages dedicated to the subject—that's more attention than the iPhone received before its launch! In this chapter, we will explore the main opinions and analyses that have been put forward. This is by no means an exhaustive list, but it will allow us to explore the great diversity of viewpoints and the broad range of areas affected by this emerging technology. Through this exploration, we will be able to identify some of the gray areas and the subjects that are rarely mentioned but may still be of importance, and we will go on to determine what aspects should be taken into account when thinking about the arrival of autonomous vehicles.

(Very) Enthusiastic advocates

For its advocates, the basic framework for thinking on the autonomous vehicle involves the (often complete and long-term) replacement of human-driven vehicles with entirely self-driving cars. In their view, the autonomous vehicle offers advantages at both individual and societal level. As consulting firm A.T. Kearney has put it:

Imagine a world with fewer accidents, fewer traffic deaths, greater energy efficiency, and lower insurance premiums. This is the world of autonomous *driving. The potential benefits are as follows: reduction in traffic accidents by 70%, lower vehicle service costs by 35%, a cut in energy consumption by 30% and a reduction in insurance liability by 15%. The estimated annual savings for the United States is $1.3 trillion, including $488 billion in savings from accident avoidance, $507 billion in increased productivity from autonomous cars, $138 billion in increased productivity from congestion avoidance and $169 billion in fuel savings.*[99]

Those savings amount to 7% of the U.S. GDP and are equivalent to the total U.S. household spending on travel. The enormity of the benefits—and above all the fact that they represent the equivalent of the country's total household spending on transportation—may be surprising to some, but it is worth trying to grasp the process by which the projected benefits—both individual and societal—will come about.

The development of fleets of electric autonomous robo-taxis will reduce the cost of transportation, largely thanks to the improved use rate of vehicles. UBS Bank goes even further than A.T. Kearney, estimating that the cost per kilometer traveled by robo-taxi will be half that of a private car (assuming that there is one passenger per vehicle in both cases).[100] If the robo-taxi trip is shared with several people, this figure will be further reduced, resulting in fares that are even lower than those of public transport in a city like London. In this context, on-demand mobility would become the preferred mode of transportation: a household that abandons a private car in favor of shared electric robo-taxi services could save 3,000 to 6,000 euros per year while traveling the same total distance. The advantages are as follows:

> **1)** The autonomous vehicle will relieve users of the task of driving and the stress that goes with it, freeing them to focus on more productive, interesting, and relaxing activities. The time spent in the vehicle will no longer be valued negatively or seen as lost time. The autonomous vehicle will be able to transport people without access to a driver's license (e.g., children, teenagers, elderly, disabled). And many parents would certainly relish the prospect of no longer having to ferry their children to and from school or to various leisure activities.

99 A.T. Kearney, *How Automakers Can Survive the Self-Driving Era*, 2016.
100 UBS Global Research, *How disruptive will a mass adoption of robotaxis be?* Q-Series, 2017.

2) The autonomous vehicle will lead to a drastic reduction in accidents. Humans are fallible. They can get distracted, become drowsy, act under the influence of alcohol or drugs, or simply make mistakes. Since 95% of accidents are attributable to human factors, we can expect to see a massive improvement in safety, as well as a significant reduction in insurance costs.

3) The autonomous vehicle will lead to an increase in the capacity of road infrastructure. Since the vehicle would be more reliable, thanks to much faster reaction times compared to the human brain, the inter-vehicle distance could be reduced. The throughput of a given type of infrastructure (e.g., a highway with three lanes in each direction) could be increased and recurring congestion could be reduced. In addition, non-recurring congestion, some of which is related to accidents, will decrease along with the drop in frequency of the number of accidents. The optimized use of space, the scarcest resource in cities, will be further improved by the use of autonomous vehicles as robo-taxis: in the long term, the private car will disappear, along with its enormous parking needs.

4) The autonomous vehicle will lead to a reduction in negative environmental impacts. The technology that controls self-driving cars is immune to nervousness or excessive pride: an autonomous vehicle will not rev the engine and start speeding just to prove itself to a bad driver who has overtaken it dangerously. The result is a smooth, seamless drive, which reduces emissions and fuel consumption. Moreover, electric engines and autonomous vehicles are mutually supportive (vehicles can recharge themselves), so local emissions could be eliminated[101] with the switch to electric engines.

Unsurprisingly then, with all these advantages, expectations and economic prospects are high. Morgan Stanley estimates the size of the autonomous mobility market at $10 trillion annually (1 billion vehicles in use x 10,000 miles per year x $1 in potential revenue per mile).[102] The investments aimed at carving out a place in this market are also considerable, in the tens of billions of dollars.

---

101 With the exception of particles resulting from the contact between the tire and the road surface and braking for local pollution, subject to "clean" power generation for global pollution.

102 "Morgan Stanley analyst on shared ownership, autonomous vehicles," *Automotive News*, January 20, 2016, www.autonews.com/article/20160120/VIDEO/301209947/morgan-stanley-analyst-on-shared-ownership-autonomous-vehicles.

With such sums at stake, the forecasts for the arrival of the autonomous vehicle are predictably optimistic, even highly so.[103] This attitude is exemplified by the "EASCY, Five trends transforming the Automotive Industry" report released by PwC.[104] It predicts a rapid transformation of the mobility sector, based on five "megatrends"[105]: electrified, autonomous, shared, connected, and yearly updated vehicles. The press kit states: "By 2030, 138 million fewer cars[106] [than today] will be needed in Europe and the United States. By 2030, more than one in three miles driven could already involve sharing concepts. 40% of the mileage driven in Europe could be covered by autonomous vehicles, and 55% of all new car sales in Europe may be fully electrified by 2030."

We can only admire the courage of a firm that ignores the advice of the futurist Nick Bostrom: "Two decades is a sweet spot for prognosticators of radical change... [It is] close to the typical duration remaining of a forecaster's career, bounding the reputational risk of a bold prediction."[107]

Naysayers who will Argue Every Inch of the Way

On the whole, skeptics accept the basic framework outlined by advocates of the autonomous vehicle and then go on to dispute their conclusions, point by point. The Canadian researcher Todd Litman is a fairly typical member of this group.[108]

The arguments put forward focus on six main points.

---

103 See Aymar, "Les prévisions des grands spécialistes pour la voiture autonome," *voiture-autonome.net*, October 10, 2015, www.voiture-autonome.net/economie/previsions-grands-specialistes-voiture-autonome-210.html.

104 The press release for the EASCY report by PwC (pwc.com, January 2018) states: "Our purpose is to build trust in society."

105 Megatrends that can be found in the Boston Consulting Group's partnership with players in the Île-de-France region, which has nevertheless taken the precaution of asking questions and exploring the necessary institutional changes: "Mobility Nation. Comment l'Île-de-France peut-elle réinventer sa mobilité d'ici à 2030. 24 propositions pour Paris 2024," bcg.com, November 6, 2017, www.bcg.com/fr-fr/d/press/6november-france-press-release-175580.

106 The total car fleet dropped from 280 to 200 million units in Europe and from 270 to 212 million in the United States.

107 Nick Bostrom, *Superintelligence: Paths, Dangers, Strategies*, Oxford University Press, Oxford, 2014.

108 See, for example, Todd Litman, *Autonomous Vehicles Implementation Predictions*, Victoria Transport Policy Institute, 2016.

1) Autonomous vehicles will be expensive at first. Before they can be sold in the millions, autonomous vehicles will have to be sold by the thousand, and then by the tens of thousands. However, if the price is passed on to potential buyers, there is a risk of choking the market or even killing it. On the other hand, financing by manufacturers requires high amounts of capital and comes with its own limitations.

2) It will also be necessary to demonstrate that there will be no need for a proliferation of roadside equipment, sensors, and other devices, particularly in the scenario where the autonomous vehicle is owned.

3) In the scenario where vehicles are used as a service, drivers perform other useful tasks (explicitly or implicitly) in addition to driving: e.g., playing the role of a trusted person on buses, capable of managing arguments or avoiding altercations by their mere presence, and providing baggage management and assistance to people with mobility challenges, to helpful taxi drivers. New ways of performing these tasks will need to be found.

4) The use of the time freed up by not having to drive depends on how the autonomous vehicles are introduced. Time will be maximized if you own a vehicle where you travel alone or with people you know, but will have less of an impact if you travel in a taxi shared with strangers (like UberPOOL). So the time-freeing benefits of the autonomous vehicle may be limited when the vehicle is shared.

5) While today's accidents are mainly caused by human error, tomorrow's accidents (and their severity) may have different causes related to the limitations of artificial intelligence (e.g., handling a situation not encountered previously) and system security problems (e.g., technical failures, hacking, cybercrime, and cyberterrorism). In addition, the severity of accidents may be exacerbated by the behavior of overconfident users (e.g., not wearing a seatbelt). Thus many people question what the real impact of the autonomous vehicle on safety will be.

6) Finally, the seductive power the autonomous vehicle holds over public authorities may distract them from their current policies in favor of pedestrians and cyclists, in terms of deterring car usage or developing efficient public transport systems.

Canaries in the Coalmine: Shifting the Debate to Other Scenarios

Many people are sounding the alarm over autonomous vehicles. We will cover three specific groups.

1) The first comes from the world of urban planning.

Their warning centers on the effects this new mode of travel will have on levels of mobility and the urban fabric. In Robin Chase's view[109], the argument that the individual autonomous vehicle would merely replace the human-driven vehicle is flawed: if the autonomous vehicle leads to a decrease in the overall cost[110] of travel, its arrival and subsequent widespread adoption would lead to more travel, more congestion, and more urban sprawl, as we have seen in the transportation revolutions of the past.[111]

For shared robo-taxis, the verdict is more favorable (see the box below). On the one hand, there would be a definitive attraction for cities where personal cars would disappear, and with them the need for street parking as they are used only 5% of the time. On the other hand, there is a (measured) risk that a less expensive mobility offer would lead to an increase in traffic.

According to Janette Sadik-Khan, the former commissioner of the New York City Department of Transportation (where she led the transformation of New York City streets into more pedestrian and cyclist-friendly places): "There are some exciting possibilities with autonomous vehicles but we need to remember what makes a great city, and that's really about the people, not the cars."[112]

---

109 Robin Chase, "Self-driving cars will improve our cities, if they don't ruin them," *Wired*, October 8, 2016, www.wired.com/2016/08/self-driving-cars-will-improve-our-cities-if-they-dont-ruin-them.

110 A trip involves several different costs, in terms of time, money, fatigue, etc. The weighted sum of these costs is called the "generalized cost." In developed countries and for commuting, the cost of time outweighs all others, and we will expand on this subject in Chapter 8 on "Uses."

111 The transition from walking to public transportation, from public transportation to the car, and so on.

112 "There are some exciting possibilities with autonomous vehicles but I think we need to remember what makes a great city, and that's really about the people, not the cars." Janette Sadik-Khan, "Street Wars 2035: can cyclists and driverless cars ever co-exist?" *The Guardian*, November 25, 2018, www.theguardian.com/cities/2017/jun/14/street-wars-2035-cyclists-driverless-cars-autonomous-vehicles.

In both these cases, there is real concern about the image of the autonomous vehicle as a "technical object" rather than a tool that can benefit the city and the people in it.

2) The second group could be classified as "transmitters of history," made up of people or institutions who have observed the path of innovation. Among them, David A. Mindell, a professor at MIT[113], notes that full autonomy has previously been pursued in submarine and aircraft technology, as well as in the domain of space exploration with the Apollo program, and that in all cases, including the moon landing, there has been an interest in man-made operations that are supported by sophisticated technical systems. In his view, full autonomy is a twentieth-century narrative (the old world). "We need to rethink the notion of progress, not as progress toward full autonomy, but as progress toward trusted, transparent, reliable, safe autonomy that is fully interactive: The car does what I want it to do, and only when I want it to do it."[114] He concludes by predicting that the fully autonomous vehicle will not be the most successful type of vehicle, for technical and societal reasons.[115]

3) The third group is an eclectic mix of people from the worlds of science, political philosophy, citizens, and law, all of whom are concerned about the disappearance privacy due to the constant collection of data. Their main concern centers around the collection and potential exploitation of the digital traces we leave behind us, and the autonomous vehicle represents an additional way of capturing our every move. Here, without any claim to exhaustiveness, we could mention: the novelist Dave Eggers, who has taken Google's idea of widespread transparency in society to its logical extreme[116]; the essayists Marc Dugain and Christophe Labbé[117]; scientists,

113 Massachusetts Institute of Technology

114 David A. Mindell, *Our Robots, Ourselves*, Viking Books, New York, 2015. Interview in *MIT News*, "Should cars be fully driverless? No," October 13, 2015.

115 It is fair to say that many of the comments in response to this interview expressed disagreement, comparing the situation to cases where highly experienced professional pilots are (or may be) at the controls, to instances involving cars, where foolish, drunk, or phone-fixated drivers are behind the wheel.

116 Dave Eggers, *The Circle*, Knopf, San Francisco, 2013.

117 Christophe Labbé, *L'homme nu, la dictature invisible du numérique*, Plon, Paris, 2016.

including Yves-Alexandre de Montjoye, who has shown that it is possible to discover the identity of a single individual by collecting just four of their credit card transactions in specific places on specific days[118]; and lawyers such as Adrien Basdevant and Jean-Pierre Mignard, who advocate the strict regulation of Big Data.[119]

Also to be filed under this category are a number of incidents that speak for themselves: for example, the hacking of Uber's database and the use of millions of Facebook users' data by a political intelligence agency. In short, we cannot simply rejoice at the arrival of connected vehicles[120]; we also need to know with whom they are communicating, how they are allowed to use the information, and if we can use this somewhat outdated term, with what duty of care.

---

118 Yves-Alexandre de Montjoye et al., "Unique in the shopping mall: On the reidentifiability of credit card metadata," *Science*, Vol. 347, No. 6 221, 2015.

119 Adrien Basdevant and Jean-Pierre Mignard, *L'empire des données, Essai sur la société, les algorithmes et la loi*, Don Quichotte, Paris, 2018.

120 The emergency calling system, which has provided strong motivation for introducing connected vehicles in Europe, is undoubtedly a useful feature.

The Use of On-Demand Transportation in Major U.S. Cities

Autonomous fleets of vehicles providing on-demand transport are an alternative to the urban use of the individual car that is most commonly promoted today. It is therefore worth analyzing the mobility of passengers who currently use a similar solution, namely private on-demand car services (i.e., taxis and ride-sourcing services like Uber and Lyft) used widely in major U.S. cities. Matthew Wigginton Conway, Deborah Salon, and David King recently did just that by analyzing the results of the 2017 U.S. National Transportation Survey[121] and comparing them to the 2009 survey.[122]

---

121 National Household Travel Survey (NHTS)

122 Matthew Wigginton Conway, Deborah Salon, and David A. King, *Trends in Taxi Use and the Advent of Ridehailing, 1995-2017: Evidence from the U.S. National Household Travel Survey,* School of Geographical Sciences and Urban Planning, Arizona State University, 2018.

The first observation: the market share of individual driver-operated passenger transportation services more than doubled, from 0.19% in 2009 (when Transportation Network Companies [TNCs] like Uber and Lyft did not exist) to 0.5% in 2017. This figure is low in absolute terms, but the penetration rate of the TNC vehicles is still high, especially considering that these services are so recent: 10% of Americans aged 16 and over use them at least once in a given month.

The second observation: these services are mainly used in large urban areas, and particularly in the densest parts of those areas.

The third observation: the higher a household's income, the more likely it will be to use TNCs (see Table 4), which is not surprising given the cost of this means of transport, which is still relatively high. However, this distribution of use is different from that observed for taxis,

	Percentage who have used ridehailing in the last 30 days	Average number of ridehailing trips in the last 30 days for ridehailing users
Race		
White	9.59% ± 0.49	4.30 ± 0.21
Black or African American	8.55% ± 0.75	5.02 ± 0.68
Asian	15.08% ± 1.47	4.72 ± 0.57
Household Income		
Less than $25,000	5.89% ± 0.78	4.53 ± 0.53
$25,000–$49,999	6.14% ± 0.68	4.17 ± 0.40
$50,000–$99,999	8.71% ± 0.54	4.39 ± 0.34
$100,000 or more	6.55% ± 1.08	4.47 ± 0.30
Age		
16–19	6.41% ± 0.58	4.32 ± 0.94
20–29	18.76% ± 1.27	4.92 ± 0.39
30–39	16.98% ± 1.21	4.67 ± 0.36
40–49	9.48% ± 0.44	3.75 ± 0.24
50–59	6.73% ± 0.45	4.04 ± 0.50
60–69	3.64% ± 0.27	3.29 ± 0.51
70–79	1.91% ± 0.25	3.02 ± 0.52
80+	1.21% ± 0.32	2.88 ± 0.85
Sex		
Male	10.50% ± 0.54	4.49 ± 0.26
Female	9.13% ± 0.40	4.32 ± 0.12

Table 4. Use of TNCs by demographic segment

Note: The data in the rows relate to ethnicity, income, age and sex; the columns include the percentage of people who have used a TNC in the past month and the average number of times per user. The margins of error correspond to a confidence interval of 95%. Source: Matthew Wigginton Conway et al., op. cit.

which are used by the richest and poorest households, and less by middle-income households. This is due to the fact that low-income households are less likely to possess their own vehicle. When they need to travel, they will occasionally use a taxi. However, they do not always have access to TNCs, which require bank accounts and smartphones. The richest households are less price-sensitive than others. They will be less reluctant to use a vehicle for hire if it is more convenient than using their own. Regardless of their income, households that use TNCs do so in a similar way, making on average four to five trips per month.

The fourth observation: users of private on-demand car services are multimodal. When at least one journey is made by taxi or TNC on a given day, the modal share of private car services during that day is only 45%. That same figure is 93% for privately owned cars. This data can be easily understood: someone who drives to work by car will certainly return home by car, whereas an individual who takes a private hire vehicle in the morning may well take public transport, walk, use a bicycle, or get a ride with a coworker to get home at night.

The analysis of the results of the U.S. National Household Travel Survey does not allow us to determine the impact of the growth of TNCs on other modes of transport, but a study based on a survey conducted in seven major U.S. cities (Boston, Chicago, Los Angeles, New York, San Francisco/the Bay Area, Seattle, and Washington, D.C.) by Regina R. Clewlow and Gouri S. Mishra provides some additional answers.[123]

It shows that the use of ride-sourcing services causes a greater disruption in the use of alternatives to the car than in the ownership and use of a personal vehicle. Users of car-sharing systems have adopted on-demand transportation en masse (65%), and some of them (23%) have even canceled their car-sharing membership. Personal vehicle ownership rates in users of TNCs who also use public

123 Regina R. Clewlow, Gouri Shankar Mishra, *Disruptive Transportation:The Adoption, Utilization, and Impacts of Ride-Hailing in the United States*, Institute of Transportation Studies, University of California, Research Report UCD ITS RR 17-07, October 2017.

transportation are higher than those who only use public transit (52% vs. 46%). Among those who never use public transportation, there is no difference in personal vehicle ownership between those who only use their personal vehicle and those who use their personal vehicle as well as TNCs. However, the meaning of these overall comparisons is up for debate, since customers of ride-sourcing services are on average more affluent. A more individual approach ("Have you given up a car since you became a customer of these services?") shows that 9% of respondents have given up at least one car since they started using TNCs. This proportion is very low among occasional customers and higher among the most regular customers (24% for daily users). Overall, 29% of respondents indicated that they use their car less than they did before.

Regarding the use of public transportation, the authors estimate that users of TNCs have used buses less (6% less), streetcars less (3% less), and commuter trains a little more (3% more) since they have started using these services, resulting in an overall decline in the number of public transport users. When asked what they would do if private hire vehicles were not available, 21% of respondents indicated that they would not travel, 24% would walk or cycle, 15% would use public transportation, 18% would carpool, 21% would drive by car, and 1% would use a taxi. What we can take from this is that the loss of customers from "virtuous" modes of transportation is not insignificant, and that the net impact of these services (less the traffic caused by personal cars) is probably an increase in traffic in the cities studied.

These results do not completely dismiss TNCs, but they do show that they can lead to an unexpected outcome for advocates of sharing services—i.e., more traffic. Another study carried out in the Île-de-France region produced a different result: the use of Uber had a negligible impact on traffic in the Paris region, as the traffic generated by drivers using the app effectively canceled out the reduction in traffic from users who would otherwise have driven themselves.[124] However, the context in France is not directly comparable to that of the U.S. In France, only services with professional drivers are allowed.

124 6t-bureau de recherche, IAU, LVMT, Jean-Pierre Orfeuil, *L'impact du service Uber sur l'utilisation de la voiture en Île-de-France*, 2018, p. 109.

The Average Joe: Wary

It's a shame to consider the potential users last, but this unfortunately reflects the position they tend to occupy in debates. Despite being highly safety conscious on average, potential users express very little appetite for autonomous vehicles and a general distrust of these machines from a safety point of view. These aspects come up consistently in surveys conducted in the major countries of the world[125], in the United States or in France, by consulting firms,[126] polling firms[127], and even by MIT.[128] Most people say that they would not be prepared to use a fully autonomous car, as they do not trust the vehicle's ability to make the right decisions. For example, MIT research shows that in 2017, only 13% of Americans were in favor of the idea of traveling in a fully autonomous vehicle (compared to 24% in the previous year).

These findings are markedly different from the results of surveys carried out by Transdev[129] among passengers of its autonomous vehicles. In that study, the level of satisfaction was consistently very high, with recommendation rates of over 95%. How can this difference be explained? We can formulate two hypotheses here. The first is related to the system operated

125 With China coming top in terms of acceptance, however.

126 Deloitte, *What's ahead for fully autonomous driving? Consumer opinions on advanced technology*, Deloitte Global Automotive Consumer Study, 2017; J.D. Power, *Hands Off? Not Quite. Consumers Fear Technology Failures with Autonomous Vehicles*, 2017.

127 Opinionway, *Sondage Automobile innovation*, 2017.

128 Hillary Abraham H et al., "Consumer Interest in Automation: Preliminary Observations Exploring a Year's Change," *Agelab*, Massachusetts Institute of Technology, White paper 2017-2.

129 Transdev is a world leader in public and private mobility services. One of the authors of this book is employed by Transdev (www.transdev.com). Transdev has transported more than 2 million passengers in its autonomous shuttles (2018 figure, combined total).

by Transdev: the speeds are low and the use cases are relatively simple (on a reserved lane or in moderate traffic, or a limited number of lanes). These conditions are unlikely to generate major concerns. The second hypothesis is that the studies mentioned above were conducted with people who had never used an autonomous vehicle. Is it possible, in such circumstances, to get a reliable idea of whether people would trust these vehicles to transport them in the future? Nothing could be less certain. As Steve Jobs said, "People don't know what they want until you show it to them."[130] It's an apt observation for something as disruptive as the autonomous vehicle.

130 Steve Jobs, "Sometimes people don't know what they want until you show it to them," *The Times*, August 26, 2011, www.thetimes.co.uk/article/sometimes-people-dont-know-what-they-want-until-you-show-it-to-them-jsqmpd9nzcn.

Chapter 5

TRUST: Five Dimensions for Analyzing the Arrival of Autonomous Vehicles

Although the first forays into autonomous vehicles happened some time ago[131], the race to autonomy is now certainly well underway. These advances bring with them a wealth of promises, but also many reservations and even warnings, as we have already seen.

Our aim here is not to offer yet another prediction, but instead to gain a better understanding of the issue by analyzing it through the five perspectives of the "TRUST" framework. This approach will enable us to cover the main challenges posed by the arrival of autonomous vehicles in various regions, as briefly outlined in this chapter.

The five factors of TRUST are:

- **Technology.** What kind of technology is involved? What are the costs? And for what benefits?
- **Regulation.** Depending on the challenges the technology poses, public authorities will need to introduce regulations, if only to allow the use of autonomous vehicles, since they are not currently authorized by traffic codes.
- **Use.** If the technology works and there are regulations that allow autonomous vehicles to operate, what benefits will there be in terms of motorized transportation? Then what will their potential uses be, and which of them will be most attractive to people?

131 These tests were conducted in the last century. As part of the Prometheus/Ertico programs, vehicles without human drivers traveled over 1,000 km (621 miles) under real traffic conditions on the streets of Paris, and a Mercedes made the Munich-Copenhagen journey, allowing itself to overtake other vehicles and peak at 170 km/h (106 mph), on the German Autobahn in 1994 and 1995, before Google and Uber existed.

- **Systems and services.** Individual preferences are just one part of a bigger picture. Autonomous vehicles are part of complex, multimodal systems involving several different players (e.g., local authorities, transportation companies, and manufacturers) that will influence the services that are actually available when up and running. But how?
- **Territories/regions.** If the autonomous vehicle becomes the predominant mode of transport, what impact will it have on the cities and metropolitan areas of the future?

TRUST is more than just an acronym. The question of trust arises at every step in the social construction of the AV. Firstly, trust in technology: at the very least, the autonomous vehicle must be as safe as a car driven by humans, and it will most likely need to have superior capabilities if it is to be approved. Then, trust in its promised benefits (aside from improved safety), such as smoother traffic and a more sustainable, peaceful, and inclusive city. Lastly, trust in the operators, who will have access to an even greater wealth of personal data than is already gathered online today, as well as urban data, which will in turn require procedures for gaining informed consent that are seemingly impossible and wholly unprecedented.

The question of trust is nothing new, either in the analysis of societies or the way they function. For many scientists, it is part and parcel of the hominization process. As they see it, the human's ability to collaborate with other humans they do not know (i.e., beyond the close circles of their family or friendship group) implies a certain degree of trust and forms the basis for a distinction between humans and other animals, helping to explain humanity's dominance over other species. For example, currency (such as banknotes) was not invented yesterday, and its role in generating trade flow has been long acknowledged. Closer to the present, major authors have played a crucial role with respect to trust: for example, the State and its monopoly of legitimate violence for Hobbes, in response to the uncertainty and widespread defiance associated with the state of nature; trust as the cement binding societies that would otherwise break apart, as argued by Simmel or Durkheim; and trust as a factor of economic development, as postulated by Arrow.

A number of factors would seem to suggest that the issue of trust is even more relevant today:

Betrayed expectations, such as progress that also comes with a downside, or large respected organizations that display dubious behavior (e.g., banks and their derivatives, car manufacturers and their emissions testing, and Silicon Valley giants and their use of our personal data).

The context of globalization, which requires us to not only place our trust in strangers, but in strangers who operate in an environment that is not regulated by the institutions we are currently familiar with, and in technologies whose operation cannot be easily checked by a group of experts, let alone the average person.

Citizens who are better educated, more inclined to follow their own path, and much more attentive to the risks that affect them, their environment, and the planet.

Lastly, and with regard to the autonomous vehicle in particular, the feeling that its widespread use could profoundly transform one of the few remaining "common goods,"—i.e., public space, which is also a place of cultural expression that can embody a society's way of life.

The need to create an atmosphere of trust around innovative technology is a rule, not an exception. This rule applies to all innovators. Platforms such as Airbnb or BlaBlaCar have found an original way of generating trust: dual rating systems that allow service providers and users to rate each other. However, the autonomous vehicle is exceptional in that it will operate in the public sphere, which means that its use will affect all citizens, whether they themselves are users or not. In a way, the "customer" of the autonomous vehicle is as much the city as the person who buys or uses the vehicle, and as such, it is up to the city to invent the means of building trust.

Chapter 6
Technology

Sensors and algorithms will have to collect information, plan, and take action.

Like living beings, autonomous vehicles use information to make decisions they then act upon. They function by coordinating three primary actions: collecting and understanding data ("sensing"), planning, and taking action.[132] We will now outline these three actions and the technologies involved, then the estimated private and public costs and benefits that could be gained from their use, before going on to discuss the outstanding issues and remaining gray areas that should be on the agenda for the next ten years.

Sense

For vehicles, this means collecting information on all the variables necessary for them to function properly—including road markings, road signs, pedestrians, cyclists and other users of personal transportation devices, other vehicles, and obstacles on the road.

Several types of sensors are used, each with its own characteristics.

- **Cameras:** They make it possible for the vehicle to "see" things at a distance, providing rich data at low cost. However, the quality of sensing drops in low-light or backlit conditions.
- **Lidar sensors[133]:** They enable the determination of the distance

132 These are the three primary actions performed by robots (Sense - Plan - Act). The logical sequence is not just a matter of Sensing > Planning > Acting. For example, an action can first be planned and then executed, while being readjusted or adapted according to the data collected during execution.
133 An acronym for Light Detection and Ranging.

between the vehicle and the objects in its surrounding environment. Lidar sensors emit pulses of light and can be used to measure distances based on the time it takes for each pulse to reach an obstacle and be reflected back to the sensor. Lidar sensors capture data at shorter distances than cameras and are much more expensive. They are also limited in that they do not function properly on some less reflective materials.

- **Radar sensors:** Like lidar sensors, they emit a signal. Distances can be calculated based on the time taken for the signal to reach an object and be reflected back. Radar technology uses radio waves instead of light beams. Radar sensors function at an even shorter distance than lidar sensors and are even more limited in terms of the materials they can accurately detect (e.g., metal objects are clearly identified because they reflect radio waves, but others are less detectable). However, compared to lidar, radar sensors are much cheaper.
- **Ultrasonic sensors:** They operate on the same principle as lidar and radar sensors, but using high-frequency sound pulses. They only provide accurate information at short distances (less than ten meters), but have the advantage of being inexpensive.
- **Infrared cameras:** They capture images just like conventional cameras, but in longer wavelengths than visible light. At normal temperature, objects spontaneously emit infrared radiation. The wavelength of this radiation varies depending on the object's temperature[134], which allows the cameras to detect objects or people (such as pedestrians and cyclists), even in darkness, thanks to the difference in heat between them and their surroundings.
- **GPS:** These systems are used to locate the vehicle in space using satellite signals. In order to function properly, they need to receive signals from a sufficient number of satellites, which is not always possible in urban environments where buildings can block signals.
- **Inertial unit:** This system is able to measure the vehicle's linear acceleration (via an accelerometer) and angular velocity (via a gyroscope), which makes it possible to determine its movement. One advantage of inertial measurement units is that they do not need

134 According to Wien's Displacement Law.

an external signal in order to function. However, they are limited in that they can drift over time due to error accumulation. They are a useful complement to GPS systems, especially when the GPS is not receiving a sufficient satellite signal (e.g., while going through a tunnel or under a bridge).

All of these sensors are positioned on the vehicle according to the role they play: front, side, or rear obstacle detectors, road marking or positioning detectors, etc. There are two reasons for using a range of sensors. One, during normal operation, each one has its strengths and weaknesses, and the data that they capture simultaneously can be used to get the clearest possible picture of the vehicle's external surroundings. A camera, for example, could be used in combination with a lidar sensor to recognize pedestrians nearby (because camera image processing is used to recognize shapes) and determine their distance and speed (using lidar information). Two, the overlapping functions of the sensors enable the system to work in degraded mode if one of the sensors fails, which can always happen.

The data collected by a vehicle regarding its real-time environment complements those saved in its memory (in particular, 3D mapping elements, which include a level of accuracy and quantity of information that goes beyond the maps currently used by human drivers). Other additional information may be provided from the environment surrounding the vehicle. This is the goal of V2I communication (vehicle to infrastructure) and V2V communication (vehicle to vehicle), which are known collectively as V2X. This input is interesting for at least two reasons. First, it makes it possible to communicate any information to the autonomous vehicle that it does not have access to, such as an accident that has occurred further along its route, the presence of ice on a bridge that needs to be crossed, or the arrival of a truck at an intersection with poor visibility, and so on. Second, the fact that the same information can be communicated or shared between several vehicles in real time means that it can be used cooperatively. For example, a line of autonomous vehicles that have stopped at a traffic light could all start moving again simultaneously, rather than in phases. These communications face a number of challenges, which will affect their future large-scale deployment. The four key ones are: standardization, performance (guaranteed levels of throughput and latency), costs, and cybersecurity.

Plan

For the next action, planning, the data collected is translated by a number of algorithms into information[135] that is then used as the basis for decision-making (e.g., the trajectory to be followed, acceleration, direction of the wheels, deceleration, activation of turn signals). Two types of algorithms are used: rule-based and machine-learning. In the first, the machine learns a set of rules: "if condition X is met, then execute Y, otherwise execute Z," etc. These algorithms work well in certain cases (e.g., "if the light is red, then the vehicle must stop"), but less so in other situations when the rules are numerous, to the point where they are almost impossible to formalize and consequently to program. This is typically the case regarding form recognition: which set of rules makes it possible to teach the machine to recognize a pedestrian captured on video?

Machine-learning algorithms work without rules, instead learning from examples (or more rarely alone). By showing large quantities of images with and without pedestrians and telling it which ones include pedestrians and which do not, the machine learns and creates generalizations by itself, which will enable it to independently recognize a pedestrian in an image that it will then analyze. This is no longer a deterministic process but a statistical one: the software evaluates the probability that a pedestrian is in the image. If this probability is high, above a certain threshold, the machine will then act on the assumption that there is a pedestrian. And conversely, when the probability is low, the machine will do the opposite. The more efficient the algorithm is and the larger the learning data set, the more accurate the probabilistic calculation will be.

These algorithms have three main purposes: route planning, localization, and the actual driving.

Route Planning and Navigation

Calculations for this activity rely on rule-based algorithms, which consider

135 We distinguish here between data, which is raw, and information, which has been interpreted.

all possible route options, rank them, and select the best ones according to the criteria selected. These algorithms, even if they are constantly being improved, are several decades old. Calculation times of approximately one second are sufficient. They are widely used today in driver-assistance systems on all vehicles equipped with GPS.

Localization

The purpose of localization is to accurately determine the location of a vehicle. Traditional GPS systems are not sufficiently accurate to perform this task on their own, so several methods are used to get around this difficulty, including real-time mapping of the environment by the autonomous vehicle and the recognition of fixed and identifiable landmarks (whose exact positions are known), which are then recorded in the vehicle's onboard 3D map.

Driving

This controls what the vehicle will do in the next few seconds. Onboard data and data from sensors is used to work out the view ahead and to create an understanding of the physical layout of the environment around the vehicle and how it changes in the short term, with the aim of defining what the autonomous vehicle must do (i.e., the precise route to be followed, acceleration, direction of the wheels, deceleration, activation of turn signals, etc.). The average reaction time required ranges from one-tenth of a second to a few seconds, depending on the vehicle's speed and proximity to the various elements in sight. In that time, onboard computers must perform all of the following tasks:[136]

1) **Create an occupancy grid**, a 3D digital model of the environment surrounding the vehicle, which is created in real time.

2) **Interpret this grid**, i.e., identify the objects that are nearby, and whether they are fixed (e.g., a curb) or moving (e.g., another

136 See Hod Lipson and Merla Kurman, *Driverless*, MIT Press, Cambridge (MA), 2017.

vehicle, a pedestrian, or a dog).

3) Assign a cone of uncertainty to each object, i.e., define the range of positions that this object could occupy within the next few seconds and continuously update this forecast.

4) Define a trajectory depending on this "predictive mapping," while taking into account the programmed destination, the highway code, and vehicle dynamics.

Tasks 2 and 3 pose the most complex problems, using both types of algorithms mentioned above to resolve them. In complex environments (a city street, for example), a rule-based method alone is effectively useless when confronted with the multitude of situations and forms that need to be recognized. There is another element of complexity: once an "object" such as a pedestrian has been detected, the cone of uncertainty generated may depend on the individual's age, whether they are on a scooter, whether they appear to be preoccupied with a cell phone, whether the person's facial expression indicates that they have seen the oncoming vehicle, or whether they are carrying a bowl or cup and are likely to approach the vehicle to ask for change, and so on. In addition, some objects are likely to lead to the presence of other objects that are not yet in sight, so for example, a bus that has just stopped is itself a fixed object, but it also indicates that some of its passengers will cross the road in the next few seconds while paying varying degrees of attention, perhaps depending on whether the bus is carrying schoolchildren or senior citizens.

Autonomous vehicles are therefore high-tech objects that may not be "good" at the outset, but they constantly improve as they accumulate experience. They remember everything. They do not suffer from memory lapses, and as soon as there is a centralized system in place, they will also remember everything that other vehicles have learned. In an "instant" world, knowledge accumulates, and for that reason AVs must be designed to be updated on a regular basis. There are limits to the amount of data that can be processed in a given time frame (to ensure the right response time)[137], but technology makes it possible, for the time being, to push

---

137 Approximately 4,000 gigabits per day, according to Denis Gingras, Director of the Vehicular Intelligence Laboratory at the Université de Sherbrooke: Denis Gingras, *Autonomous vehicles: the work done so far and some remaining challenges*, SMIV VEDECOM Conference, 2018. 4,000 gigabits is equivalent to 500 hours of Netflix.

these boundaries. And that is also why the various players developing AVs are currently doing so much real-world testing.[138] For a specific skill, the vehicle with the most accumulated data and the broadest experience of different situations will be the best[139], and therefore will have the best chance of success.

Act

Here, the aim is to convert a command from the computer system into a mechanical action (e.g., braking, acceleration, or changing the angle of the wheels) using the vehicle's actuators. Although the principles involved are old, the control technologies have evolved significantly, and now require reaction times that are measured in milliseconds. These controls work better with electric motors than with internal combustion engines, which may have a higher lag time.

Costs

If they are to be adopted, technologies that make it possible for vehicles to operate without a driver (i.e., Level 4 and 5) must be sufficiently low in cost, or else they must offer the user benefits that outweigh the additional costs. This will likely turn out to be the case in the long term, but there is still a long way to go, because the entry point comes with considerable additional costs.

The Total Cost of Ownership (TCO) for the User

Autonomous vehicles are currently much more expensive to manufacture than conventional vehicles, requiring sensors, software, onboard com-

138 These experiments are essential, but they are not sufficient to demonstrate that the autonomous vehicle is better than a human. Based on the rates of serious accidents involving human drivers and the daily distances traveled by the autonomous vehicles being tested today, Denis Gingras has shown that it would take several decades of testing to statistically demonstrate this. In addition, these players use different vehicles, and the software is constantly being updated. Denis Gingras, *op. cit.*

139 It is the best one, but it too has its failings: the range of situations encountered in the real world is infinite.

puters, and robotic components in order to perform the driving functions that are currently carried out by humans. This technology is costly: more than $100,000 for the majority of military and civilian AVs in circulation in 2013—but the price is likely to drop dramatically. Daniel Fagnant and Kara Kockelman estimate that it could cost $3,000 or less, when it has reached road-readiness.[140] In December 2017, Kevin Clark, CEO of automotive supplier Delphi, announced the figure of $5,000 (for hardware and software) by 2025.[141] This reduction in costs can be explained by the economies of scale achieved by the automotive industry, which can produce several million units of highly sophisticated components per day for a few dollars each, as well as by the elimination of various types of equipment that are geared toward human drivers: the dashboard, steering wheel, and pedals, etc. It is worth noting that $5,000 is comparable to the increase in the average selling price of new vehicles in France between 2010 and 2016: a rise of 6,000 euros.[142]

In a world where only autonomous vehicles are in circulation, the risk of accidents would be reduced (see below). Passive safety systems (e.g., strengthening the structure of the vehicle, airbags) could become superfluous, which would further reduce the production costs per unit.[143]

So from here onwards, we will assume that eventually the additional cost of purchasing an autonomous vehicle as opposed to a conventional one (if any still exist at that point) will be comparable to the difference in cost between a basic conventional vehicle without any special features and an advanced model with all the optional extras (e.g., parking assistance, speed adapters, and so on). It should be added that the real value of autonomous driving technologies is bound to vary greatly from one vehicle to another, depending on the complexity of the use cases they will have

140 Daniel J. Fagnant and Kara Kockelman, "Preparing a nation for autonomous vehicles: opportunities, barriers and policy recommendations for capitalizing on self-driven vehicles," *Transportation Research*, Part A 77, 2015, pp. 167-181.

141 Paul Lienert, "Self-driving costs could drop 90 percent by 2025, Delphi CEO says," *Reuters*, December 4, 2017, www.reuters.com/article/us-autos-delphi/self-driving-costs-could-drop-90-percent-by-2025-delphi-ceo-says-idUSKBN1DY2AC.

142 The average price of a new vehicle was 19,767 euros in 2010; it rose to 25,828 euros in 2016. Source: Xavier Chimits, "Voiture moyenne neuve 2016: la hausse du prix se poursuit," April 26, 2017, https://pro.largus.fr/actualites/voiture-moyenne-neuve-2016-la-hausse-du-prix-se-poursuit-8504398.html. Accessed June 25, 2018.

143 In the transitional phase, on the other hand, there may be an increase in the risk of accidents, particularly because drivers of traditional vehicles tend to incorrectly anticipate the behavior of autonomous vehicles (since these machines are extremely compliant with rules and regulations). This calls for the maintenance of these systems during a long transitional phase and for autonomous vehicles to be equipped with signs allowing them to be identified by other road users, be they pedestrians or "traditional" motorists.

to encounter. For example, a shuttle operating in Grand Canyon National Park would not need as many sensors or as much computing capacity as a vehicle that navigates on Broadway in New York City.

The cost of maintaining autonomous vehicles will be considered comparable to the maintenance required for conventional vehicles: the servicing and maintenance of autonomous driving technology will involve additional costs, but improvements in safety will reduce the maintenance requirements of conventional components.[144,145]

Insurance costs are expected to decrease as a result of the reduction in accidents. However, this drop could be partially offset by an increase in the cost of vehicle damage due to vandalism, which could be prevented to some extent by fitting the most expensive equipment under the bodywork of the car.

Lastly, costs related to energy consumption are also expected to fall (by around 10% according to the U.S. Department of Energy[146]) due to smoother driving.

The Potential Costs of Infrastructure Equipment

Beyond the costs of owning an AV itself, there is also the question of the cost of the infrastructure needed for them to function. In this area, the approaches differ hugely depending on the manufacturer: some of them (e.g., Waymo and GM) are developing a vehicle that will be able to drive on the existing road network, without ad hoc retrofitting or the need for continuous communication with the outside world. In this case, we can only hope that the broken and continuous lines marking the lanes on roads are well maintained. Doubts around the ability of public authorities to commit to costly infrastructure equipment policies, due to the sheer number of AV

144 However, it is also necessary to implement *low-tech* technologies, such as fluid distribution channels to clean camera lenses, and, in the case of shared autonomous vehicles, to ensure that the lenses have not been blocked by chewing gum, for example, which is common in video surveillance systems in buildings.
145 Patrick Bösch, Felix Becker, Henrick Becker, and Kay W. Axhausen, "Cost-based analysis of autonomous mobility services," *Transport Policy*, No. 64, 2018, pp. 76-91.
146 T. Stephens, J. Gonder, Y. Chen, Z. Lin, C. Liu, and D. Gohlke, *Estimated Bounds and Important Factors for Fuel Use and Consumer Costs of Connected and Automated Vehicles*, Technical Report, National Renewable Energy Laboratory, U.S. Department of Energy, Golden (CO), November 2016.

players involved, are seen as outweighing the potential benefits. In addition, not being permanently connected reduces the risk of hacking.[147] Other manufacturers (mainly in Europe) envisage their vehicles communicating not only with each other, but also with the road infrastructure and its equipment, such as traffic lights.

The Expected Benefits

There are many expected benefits for the AV user: the ability to travel by car without a permit, to fill the time that would have been spent driving with other activities, or to offer driverless transportation services, etc. The benefits expected by AV proponents, in particular those resulting from the monetization of data, are potentially enormous, and we will go into those in more detail later on.

When launching their experiments, promotors of autonomous vehicles tend to focus on the benefits to society as a whole, and mainly in relation to road capacity and road safety. It is these benefits we will be examining more closely here.

Road Network Capacity

The optimized driving of AVs could make it possible for more of them to circulate at the same time on a given section of road, thanks in particular to their Adaptive Cruise Control features, which would result in reduced intervehicular distance, compared to the usual distance needed between vehicles driven by humans.

However, it is the connectivity between vehicles, as well as the connectivity between vehicles and infrastructure, that appears to offer the most potential in terms of increased road network capacity. Using this interconnectivity, speed and trajectory are no longer optimized vehicle-by-vehicle but across the entire system. For example, Cooperative Adaptive Cruise

147 Jamie Condliffe, "Why Some Autonomous Cars Are Going to Avoid the Internet," *MIT Technology Review*, January 10, 2017, www.technologyreview.com/s/603339/why-some-autonomous-cars-are-going-to-avoid-the-internet.

Control functions enable traffic to avoid accordion effects, in the case of slowing down or accelerating along a section of the road; they also enable the adoption of the traffic speed that corresponds to the best traffic flow[148], or that will optimize traffic flow at intersections or areas where there is a reduction in the number of circulation lanes.[149]

However, studies and modeling on this subject have led to a very wide range of predictions: the estimated increase in road capacity caused by replacing human-driven vehicles with autonomous and connected vehicles ranges from 1% to more than 400%[150], with an average increase of 100%.

In addition to the positive effects of cooperative driving, there are also the positive effects of improved safety. If robotic driving is safer than human driving (see Road safety paragraph below), we can assume that the associated reduction in the number of incidents and accidents will lead to improved traffic flows, since a quarter of congestion-related events are caused by traffic incidents.

The transition to autonomy could therefore result in a major overhaul of estimated infrastructure requirements, particularly in countries where the automobile system is yet to reach maturity. In countries that are well equipped in terms of infrastructure, this transition could effectively reverse the trend when it comes to service quality: instead of a slow deterioration, users could expect constant improvements with the same infrastructure, provided the rebound effects are minor.

---

148 Meng Wang, Winnie Daamen, Serge P. Hoogendoorn, and Bart van Arem, "Connected variable speed limits control and car-following control with vehicle-infrastructure communication to resolve stop-and-go waves," *Journal of Intelligent Transportation Systems*, Vol. 20, No. 6, 2016, pp. 559-572; Steven E. Shladover, Xiao-Yun Lu, and Dongyan Su, "Impacts of Cooperative Adaptive Cruise Control on Freeway Traffic Flow," Transportation Research Record, *Journal of the Transportation Research Board*, 2012; A.A. Delisa, I.K. Nikolosa, and M. Papageorgiou, "Macroscopic traffic flow modeling with adaptive cruise control: development and numerical solution," *Computers and Mathematics with Applications* 2015.

149 J. Rios-Torres and A.A. Malikopoulos, "Impact of connected and automated vehicles on traffic flow," UT-Battelle, LLC, under contract No. DE-AC05-00OR22725 with the U.S. Department of Energy, 2016; Y.J. Zhang, A.A. Malikopoulos, and C.G. Cassandras, "Optimal control and coordination of connected and automated vehicles at urban traffic intersections," UT-Battelle, LLC under contract No. DE-AC05-00OR22725 with the U.S. Department of Energy, 2017.

150 Vincent A C van den Berg and Erik T. Verhoef, *Robot Cars and Dynamic Bottleneck Congestion: The Effects on Capacity, Value of Time and Preference Heterogeneity*, Tinbergen Institute, discussion paper, 2016.

Road Safety

The U.S. Department of Transportation has done a comprehensive study[151] of road accidents involving at least one vehicle that had to be towed due to the damage caused. The study shows that 95% of accidents were due to human factors.[152] The mistakes made are of a diverse nature, with errors related to processing information (e.g., inattention and distraction) and to decision-making (e.g., driving too fast, illegal maneuvers, and insufficient inter-vehicle distance) at the top of the ranking (see Table 5).

Critical Reason for Critical Pre-Crash Event		Number of Crashes		Weighted Percentage
		Unweighted	Weighted	
Recognition error	Inadequate surveillance	1,080	414,626	20.3%
	Internal distraction	482	218,548	10.7%
	External distraction	229	77,496	3.8%
	Inattention (i.e., daydreaming, etc.)	194	65,712	3.2%
	Other/unknown recognition error	109	51,926	2.5%
	Subtotal	2,094	828,308	40.6%
Decision error	Too fast for conditions	348	171,604	8.4%
	Too fast for curve	181	100,713	4.9%
	False assumption of other's action	260	92,583	4.5%
	Illegal maneuver	232	78,112	3.8%
	Misjudgment of gap or other's speed	212	65,221	3.2%
	Following too closely	85	30,452	1.5%
	Aggressive driving behavior	99	31,026	1.5%
	Other/unknown decision error	335	125,805	6.2%
	Subtotal	1,752	695,516	34.1%
Performance error	Overcompensation	211	100,090	4.9%
	Poor directional control	249	95,165	4.7%
	Other/unknown performance error	30	7,751	0.4%
	Panic/freezing	20	7,137	0.3%
	Subtotal	510	210,143	10.3%
Non-performance error	Sleep, actually asleep	160	65,141	3.2%
	Heart attack or other physical impairment	133	48,822	2.4%
	Other/unknown critical nonperformance	76	31,881	1.6%
	Subtotal	369	145,844	7.1%
Other/unknown driver error		371	162,132	7.9%
Total		5,096	2,041,943	100%

Table 5. Causes of accidents that occurred due to human factors (driver)

Estimates may not add up to totals due to independent rounding. Source: U.S. Department of Transportation, National Highway Traffic Safety Administration, National Motor Vehicle Crash Causation Survey, Report to Congress, 2008, p. 25.

151 U.S. Department of Transportation, National Highway Traffic Safety Administration, *National Motor Vehicle Crash Causation Survey*, Report to Congress, 2008.

152 The other 5% are due to vehicle-related (e.g., mechanical problems, worn tires) or weather-related causes (e.g., icy roads, snow, fog) in equal proportion.

The most eagerly anticipated benefit of the autonomous vehicle is the ability to avoid the mistakes that humans commonly make, and thereby a massive potential reduction in accidents. However, AVs may make other mistakes, especially in the early stages when their systems have not been sufficiently trained. For that reason, it will be important to ensure experiential feedback is as comprehensive as possible. Autonomous vehicles also carry other risks, such as those related to cybersecurity.

Still, it may be assumed that the balance will be largely positive (i.e., fewer accidents with autonomous vehicles) by employing a reverse kind of reasoning. Self-driving cars will only be allowed on the roads if they are more effective than humans—although at this stage it is impossible to say how much more effective they will have to be.

It is possible that significant improvements will be required. We know that the social acceptability of a death caused by a professional driver or a machine is lower than that of a death due to other human causes.[153] A bus or train accident generates more reactions, more media coverage, and more political decisions than a car accident, even when there are the same number of victims. When the professional driver is a computer connected to sensors, there is no reason to believe this will change. Moreover, it is less socially acceptable for the same accident to occur repeatedly, in which case a judge is likely to find negligence. Each accident involving autonomous vehicles will therefore need to be investigated to discover its causes, and the system will have to be corrected in order to prevent the entire fleet equipped with the same technology from repeating the same accident and being ordered off the road.

This approach is not a perspective shared by everyone, particularly in the United States. Some authors point out that when it comes to safety, any delay in adopting technologies that are only "a little" better at driving than humans will have a human cost in the lives that could otherwise have been saved.[154] The rapid introduction of autonomous vehicles—once they have reached the average human level of driving ability—would then be statis-

153 This is why the "shadow price" (i.e., theoretical value) of an accidental death has been twice as high for a road accident involving public transportation as for an accident involving a private vehicle, according to socioeconomic calculations in France. It is now fixed at 3 million euros (2010 value) regardless of the mode (France Stratégie, *L'évaluation socio-économique des investissements publics*, 2013).
154 Nidhi Kalra and David G. Groves, *The Enemy of Good: Estimating the Cost of Waiting for Nearly Perfect Automated Vehicles*, RAND Corporation, Santa Monica, 2017.

tically beneficial, despite the many accidents that would occur involving autonomous vehicles. The implicit assumptions are that the press will tire of reporting the accidents and that the deaths related to human-driven vehicles and to autonomous vehicles will eventually have the same value.

Unresolved Issues

The first gray area in this debate centers around the momentum of a process where the point of arrival (i.e., a world where human-driven vehicles have disappeared) may be considered better in many ways than the existing situation, but where the decades between us and that point of arrival are characterized by very high costs and low or invisible benefits. When the autonomous vehicle was climbing the hype cycle, this problem was not an issue, because the positive momentum was enough to attract investors who were willing to bet big to win the jackpot. But a different picture is now emerging, and the prospect of full autonomy seems to be moving further away.

The second gray area revolves around the validation of an algorithm based on machine learning. There are several difficulties involved.[155]

First, we will deal with the technical issues, then move onto more political concerns.

Algorithms based on machine learning work in a non-transparent way: they are "black boxes," so to speak. The process that leads the software to produce a particular result based on a given input is generally unintelligible, which makes it impossible for a third party (e.g., a certification body) to check that it is functioning properly using deductive reasoning. We therefore need to use other benchmarks for decision-making. If, as is likely to be the case at first, the use of a Level 4 vehicle is approved in a given context, what contextual classification framework can be used to avoid a proliferation of procedures or running up hundreds of millions of miles in approval testing? For example, digital simulation is one option, but can it accurately represent the real world?

155 Rick Salay, Rodrigo Queiroz and Krzysztof Czarnecki, "An Analysis of ISO 26262: Using Machine Learning Safely in Automotive Software," *ArXiv*, SAE technical paper, 2017; Philip Koopman and Michael Wagner, "Challenges in Autonomous Vehicle Testing and Validation," *SAE International Journal of Transportation Safety*, Vol. 4, No. 2016-01-0128, pp. 15-24.

The AV's probabilistic mode of functioning means that there will always be a certain rate of error. This rate can be estimated by conducting a test on a sample of data (e.g., images), which is necessarily finite and therefore limited: how well does it represent real-life situations, and the potentially infinite situations the vehicle may encounter? Finally, almost identical input data can produce very different output results, making it difficult to interpret any vehicle test, even in the most standardized environment.

Safety issues concern both algorithms and sensor data. Updating the autonomous vehicle's software (rule-based as well as machine-learning) to improve its performance could also open the door to malicious attacks on the system, particularly when the update is carried out via a wireless connection. This possibility is not theoretical: in 2015, two hackers managed to take control remotely of most of a Jeep's functions.[156] Even when not using a wireless connection, a DIY-minded owner could modify a car in their garage.

Furthermore, the AV's autonomous driving system may be led to make bad decisions, without being contaminated by a virus or other malicious software, if its sensors provide it with corrupted data. This type of security risk could occur in the presence of a GPS jammer sending false satellite information. A lidar sensor could also be deceived in the same way, by receiving false signals that correspond to false obstacles, for example.

There is yet another type of security risk, which is linked to malicious intent, but does not involve hacking into or deliberately misleading the autonomous driving system. For example, if, for geopolitical reasons, a foreign supplier of the software components that control a vehicle decides to halt all the autonomous vehicles installed with its software in another country, or even worse, decides to send them all at high speed to high-traffic areas, the country could be brought to a standstill or become vulnerable to attack. The issues of national sovereignty raised by this possibility have until now been relatively absent from the debate, even though the production of autonomous vehicles involves the collaboration of players from a number of different countries.

156 Andy Greenberg, "Hackers Remotely Kill a Jeep on the Highway – With Me in It," *Wired*, July 21, 2015, www.wired.com/2015/07/hackers-remotely-kill-jeep-highway.

In addition to these technical issues, there are also problems of a more political nature. As there are so many of them, we will simply list them at this point.

However capable it is, a certification body can only be of use if it is trusted by the public. Which body is most trusted? If the most legitimate body only represents a limited area (e.g., a city or country), how can its limited scope be reconciled with the challenges involved in the global standardization of autonomous vehicles?

Experiments in the real world cannot be conducted in the same context as those requiring informed consent, such as trials that have been developed in the medical field, where patients are fully informed, agree to participate, and can withdraw from the process at any time. What procedures could we use to get closer to that model? And if we fail to do that, what are the risks of public rejection, and how can we then reopen the dialogue?

What rates in terms of errors, breakdowns, and accidents are we collectively willing to accept? How can we determine them collectively?

How can we coordinate public feedback on the user experience when the players involved are locked in ruthless competition?

Chapter 7
Regulations

Managing the Transition from the Old World to the New

With the autonomous vehicle, we find ourselves at the beginning of a long process, in what Polish philosopher Georges-Hubert de Radkowski termed "the moment of dawn":

If there is an opportune moment for a new technology to cover up its negative side, its underside, so it can be seen in a purely positive light as a force for freedom, unencumbered by the trials and tribulations it will leave in its wake, it can only be in that moment of dawn, when the technology has not yet reshaped the old world it came from. Designed to free humankind from the constraints of the existing world, its achievements are, for a moment, unhindered by the new constraints brought about by its creation.[157]

This inaugural moment is ripe for analogy. The autonomous vehicle is widely seen as a "smartphone on wheels," and, it must be said, the temptation to draw parallels between the arrival of the autonomous vehicle and that of the Internet and the smartphone is indeed hard to resist.

Physical Space is not Virtual Space

This fact may seem an unnecessary reminder, but it is needed given the implications of the fundamental differences between these spaces. Tech

157 Georges Hubert de Radkowski, *Les jeux du désir*, Puf, Paris, 1978.

giants have generated their power and changed our lives by investing in a new virtual space that had been untouched by history, customs, and traditions, and where sites generally operate independently from the servers that connect them to the world. These giants have occupied this space and invited us to join them, while at the same time laying down their own rules and capturing and using our data, without telling us exactly how or why. In this space, the extreme rigor associated with digitization (where ambiguity and vagueness are not an option) coexists with the extreme flexibility of the interfaces that make life easier for users. Internet users from all over the world have felt right at home from the outset. There is no need for any code of conduct when these users visit the same places without interacting with one other. Despite the concentrated co-presence (i.e., being occupied by many different people at the same time) in this virtual realm, to each user the presence of any other person is often intangible; it cannot be detected by the senses, so it remains invisible.[158] In this world, the flow of traffic can be maintained by the seemingly limitless expansion of its networks.

The space we move about in on a daily basis is of a completely different nature. The physical world is finite; space is scarce and sought-after; the potential for new infrastructure is limited. There is much more inertia in the physical world than in the virtual world. In the developed world, urban planners point out that 80% of the city that will exist in twenty years has already been built. The city often tells us a story that conveys a deeper truth. The occupants of an area (e.g., inhabitants, storekeepers, and companies) have a certain actual influence (if not a legal one) over the use of the networks that serve them (e.g., parking regulations and the degree of acceptance of transit traffic). These spaces are inhabited, occupied, and used in a different way in the center of Lyon versus a suburban village of Paris or Tokyo, and even more so in Dhaka or Dakar. Although there are many rules on the use of space (for example, the highway code), they do not regulate everything, if only because some people may not comply with the rules or even be aware of them. In addition, breaches of the rules are common and are often tolerated in order to keep the system moving and functioning (e.g., crossing a centerline to overtake a double-parked vehicle).

---

158 However, the situation has changed with social networking and the increasing "horizontalization" of communication, for better or worse (online harassment that can lead to suicide, pedophile sites or the incitement of terrorism, etc.).

There is no physical sense of danger in the virtual space.[159] In the real world, any vehicle represents a potential risk[160], and the mere sight of a vehicle driving itself could be distressing, let alone the thought of getting into one. In this space, other people are present in person, with a whole range of capabilities and powers depending on whether they are pedestrians or motorists, and with very different rights depending on the place[161] and time (see the box below). When a new means of transportation first appears, the rule of the survival of the fittest begins to apply, but then rules of conduct are established and gradually become accepted. Each potential conflict situation (e.g., a pedestrian crossing the road, or two passengers passing by on a narrow sidewalk) gives rise to informal, usually non-verbal exchanges that end up coordinating each party's decisions, but two seemingly identical situations could result in different decisions. Last of all, the channels of communication in the real world are not infinitely expandable.

It would be impossible to envisage the arrival of autonomous vehicles without taking these differences into account. Historically, the deployment of new systems based around new vehicles has always taken time, even when the early adopters belonged to the upper classes and were able to override the objections of those who did not have access to these new modes of transport.

These new modes had to deal with the existing ones and operate alongside the established modes and usages, while gradually transforming the urban experience by introducing new rules, until the city itself was transformed.

159 A computer can be made unusable, and a ransom can be demanded for the return of stolen data, but a remote terminal cannot yet be blown up.
160 More than one million road deaths (mainly pedestrians) per year worldwide.
161 For example, laws and civil behavior in the United States and France currently provide better protection for pedestrians than those in Latin America.

It's Not Easy to Carve Out a Place for Yourself when You're the New Kid on the Block (Who will One Day Grow Up)

Until the mid-nineteenth century, there was only one mode of transportation available on city streets (apart from walking): the horse-drawn carriage. While this might sound like a stable system, in reality it was anything but. Firstly, the traffic of these vehicles was steadily increasing. Secondly, these vehicles were diversifying. Lighter and faster carriages such as cabriolets were emerging, while others were getting larger to provide passenger transport. These developments provoked reactions and led to new rules and expectations. Some of those reactions in France can be found in the book L'Invention de la vitesse (The Invention of Speed)[162]:

1) First, regulations. Traffic was to take precedence over the static use of space. The seventeenth century saw a succession of decrees aimed at combating the obstruction of streets by static objects (e.g., piles of wood) and the appropriation of street space by its residents (p. 25). The movement of carriages was to be facilitated: urban planning ordinances were issued to improve the straightness of existing streets and to impose a minimum width on new streets (p. 65); while speeding was prohibited (King Francis I of France issued an order prohibiting "speed racing" in the streets of Paris). Gradually, it became necessary to "share the road." The sidewalk was invented shortly before the French Revolution to separate the road for vehicle use (p. 65), and pedestrians were ordered to use these sidewalks by the Paris Prefect of Police, Gabriel Delessert (p. 113).

2) Then, spontaneous reactions. The public revolted against the carriage, "that damnable invention," which had the privilege of "crushing, milling and pulverizing citizens" (p. 67).[163] In 1790, a citizen made the following appeal: "You fickle and frivolous public, you take offense to the coats of arms that decorate carriage panels and hurt only your pride, yet you forgive the wheels that crush, maim and constantly

---

162 Christophe Studeny, *L'invention de la vitesse*, Gallimard, coll. "Bibliothèque des histoires," Paris, 1995. The text refers to the page numbers in the book.
163 Louis-Charles Fougeret de Montbron, *La capitale des Gaules, ou la nouvelle Babylone*, La Haye, 1759.

startle you," (p. 71). The perspective of a foreign visitor (in this case a Scottish man) from 1775 is still useful: "You have to run to the shelter of each doorway, for fear of being run over by the wheels of the carriages" (p. 64). The situation seemed to worsen every year. In 1837, Delphine de Girardin noted: "Walking has become impossible. Anyone strolling the streets faces the death penalty. We no longer walk, we run. Walking is a battle. There are no more passersby; only travelers" (p. 113).[164] Later, in 1883, Théodore de Banville wrote: "Crossing Rue de Rivoli at 5 o'clock is undoubtedly a more perilous undertaking than exploring Central Africa surrounded by cannibal tribes."[165] The horse-bus, on the other hand, seemed to meet with the personal approval of another novelist, as he noted:
"Personally, I go from place to place on the horse-bus. I plan my itinerary using a map of Paris and a guide to the routes and connections. I see a lot of things, more than you do. I would like it to the sensation of taking a trip around the world, because the people are so different from one street to the next."[166] Others also note their approval on ethical grounds: "The horse-bus makes distances
seem shorter; it brings together people of every class and rank. It has made travel by horse and carriage a public right." (p. 116)[167]

3) Finally, expectations. The covered passageways that were invented to allow people to window-shop in peace were doomed when the first road-widening projects took place: "The passageways are now at their peak. The street is becoming large and wide, abutting on covered porticoes,

164 Delphine de Girardin, *Chroniques parisiennes, 1836-1848*, reprint: Éditions des femmes, Paris, 1986.
165 Théodore de Banville, *Paris vécu*, Charpentier, Paris, 1883.
166 Guy de Maupassant, "Le père Mongillet," *Gil Blas*, 1885.
167 Taxile Delord, Arnould Frémy, Edmond Texier, and Maurice Alhoy, *Les petits Paris*, No. 13: "Paris-en-omnibus," Taride, Paris, 1845.

and it is taking over the sidewalk" (p. 109).[168] Haussmann explained in his project: "We need wide roads, which lead to all the railway stations, veritable hubs of the new roads of France and Europe, which establish easy transportation between all parts of the city, allowing carriages to move quickly, pedestrians to travel safely and without obstacles, thus shortening distances and saving time, that precious resource of business." Other authors quickly identified the "impacts of the infrastructure": "The wider the streets, the harder it is to pass through them. Carriages clutter up the vast road, pedestrians fill the vast sidewalks, Paris is flooded."[169] And again, these surprisingly modern-sounding words: "We are demolishing old streets because they could not accommodate the traffic of carriages. The more we demolish, the more Paris has to be expanded in both directions. The more it expands, the longer our journeys, the harder it is to walk around the city, the more necessary carriages become."[170] And "the railways are expanding the circle; a person who formerly rented a country house a few miles from Paris can now look for one much further away without spending any more time on it."[171]

168 Auguste Luchet, "Passages" in *Nouveau tableau de Paris au XIXe* siècle, Vol. 6, Paris, Mme Charles-Béchet, 1835.
169 Louis Veuillot, *Les odeurs de Paris*, Palmé, éditeur des Bollandistes, Paris, 1867.
170 Edmond About, *Paris guide*, Lacroix, Paris, 1867.
171 *Le commerce*, 1839.

It's likely that the same sequence of events will occur with autonomous vehicles. In this chapter, we will look at the new regulations that should or may be introduced in six different areas: certification, highway codes, ethics, data, financing, and the management of local public space.[172]

172 The transformation of the city will be explored in Chapter 10, Territories/Regions.

Certification

There are currently no performance standards in place for autonomous driving. Such standards should be developed to determine whether a vehicle or transportation system is capable of carrying passengers while ensuring an adequate level of safety for these passengers and other people on the roads and streets.

Who will define those standards? Which authorities will be in charge of implementing them and certifying new vehicles?

In Europe, emissions authorizations and safety permits are issued at the European Union (EU) level. It is likely that autonomous vehicles will be authorized at the same level, as this framework is helpful in ensuring fairness between domestic manufacturers and the interoperability of their devices. The EU is most likely to play the role of mediator between manufacturers, lobbies, and the majority of the public (who may be more reserved). To this end, it will probably be helpful—and in fact it is customary—to involve the European Parliament and forums for public debate in the process. Also, and most importantly, it will be necessary to find an appropriate way to involve cities and regions in the process: any defect in a vehicle or operating system will directly impact not only the vehicle and its occupants, but also the entire environment in which the vehicle operates[173], and it is the local elected representatives whom citizens will turn to in the event of such a problem, not the EU certification bodies. Furthermore, because the supranational body has very weak political legitimacy in local communities, additional intervention by states, as in other sectors[174], cannot be ruled out, especially given that the idea of the state as a protective force varies hugely from one country to the next, even within the EU.[175]

In the United States, the federal government is responsible for regulating the safety of road vehicles, their equipment, and their interstate commercial operation. States and local governments are responsible for issuing driver's

173 Here we are thinking in particular of the challenges involved in cybersecurity (e.g., how would Paris, Barcelona, London, or New York function if thousands of autonomous vehicles were blocked, bringing the city to a standstill?) and terrorism (which could be carried out remotely).

174 In the pharmaceutical industry, there are national authorities and a supranational authority.

175 The demand for a protective state is much greater in France than in the Netherlands, for example.

licenses, establishing their traffic regulations, and laying down rules on liability and insurance. Based on the first few publications of the U.S. Department of Transportation on the subject of autonomous vehicles, no change to this distribution of roles seems likely.

In any event, and in all countries with a high level of risk aversion, the certification body will have to win the trust of the public and convince people that it is possible to certify (regularly updated) machine-learning technology that is not compatible with traditional, Cartesian approaches to proof or even probabilistic approaches to risk.[176] States may wish to strengthen civil liability law and condition the use of these vehicles by engaging the responsibility of a trusted third party, one with both a reputation for excellence and sufficient financial backing. This trusted third party might be a transportation company (such as RATP, Transdev, Deutsche Bahn, or MTR), a local authority or insurer, or even, from the perspective of a smart city, a company in the energy or urban services sector.

Traffic Regulations, Street Regulations

States will also have another role to play: making changes to their highway codes. First to authorize the use of driverless vehicles, and then to lay down a series of new traffic regulations. For example, under current usual rules, a person at a pedestrian crossing without traffic lights could in principle stop an autonomous vehicle for as long as they wish. What sense would there be in spending hundreds of billions of dollars, just to afford a person (perhaps not even an adult) the right to stop a vehicle simply by standing in its way? What rights and duties will be given to the pedestrians, autonomous vehicles, and other modes of transport of the future?

Beyond these legal issues, the way the various functions of the street are organized will also have to be redefined with the arrival of autonomous vehicles. What are the priorities for developing and using sidewalk curbs? There will be many demands (e.g., charging stations, delivery areas, self-service bicycle stations, and drop-off and pick-up points). Urban space will remain limited. How can we ensure mutual respect between

176 In particular, the ISO 26262 safety standard for electrical and electronic systems in road vehicles.

different forms of transport? How can we ensure the safety of the most vulnerable citizens? How can we prevent small autonomous delivery vehicles from proliferating uncontrollably and reversing the anticipated benefits of having fewer cars on the road? The relationship between the different people affected on the streets and roads will be one of the priority topics to be addressed, if we are to guarantee the safety of all involved and ensure that the system is legally sound.

Ethics

Although autonomous vehicles offer significant benefits in terms of safety, not all accidents can be prevented, especially those where no error is committed. For example, if a pedestrian, obscured by a parked truck, decides to run across the road, a nearby autonomous vehicle may not be far enough away to stop in time, and the incident could result in a fatality.

There is also another scenario we can consider. The vehicle could choose to brake to the best of its ability and manage to avoid the pedestrian, but it could hurt other potential victims instead—e.g., those traveling inside the autonomous vehicle, who could be jolted violently by an emergency stop, especially if they are standing, as is common with public transportation. This dilemma is known as the Trolley Problem, and it was originally devised by British philosopher Philippa Ruth Foot in 1967: an out-of-control streetcar that is unable to stop is heading toward five workers on a track. These five workers will be killed unless the streetcar driver chooses to divert it to another track, where there is only one person. What decision should be made: would you choose to kill one person to save five?

This question is no longer theoretical when it comes to the autonomous vehicle, as we have just seen in the case of the inattentive pedestrian. The first company to announce its decision on this issue was Mercedes-Benz: it stated that it would always prioritize the passengers in its vehicles over third parties outside.[177] However, the car manufacturer had to quickly backtrack in response to criticism. How can it weigh up the value of the lives of its customers against the lives of other road users? There is obviously

177 David Z. Morris, "Mercedes-Benz's self-driving cars would choose passenger lives over bystanders," *Fortune*, October 15, 2016, fortune.com/2016/10/15/mercedes-self-driving-car-ethics.

no "right" solution to the Trolley Problem. The variations are endless: is it better to save a young pregnant woman than an old man? What about two older men? A hundred older men? And so on.

Germany is the first country to have issued ethical recommendations for autonomous vehicles.[178] Its guidelines indicate that when an accident is unavoidable, the protection of human life must be given priority over material damage and animal life, and that in the event of unavoidable human losses, the choice of one life over another cannot be based on the individual characteristics of the people concerned (i.e., age, gender, mental or physical makeup). The decision will have to be made using other contextual factors (such as the unpredictable behavior of an individual) under conditions that have yet to be defined. That responsibility has been handed to an independent public agency.

Data

Autonomous vehicles will or may be able to collect and generate large amounts of data. There are four types:

1) Data relating to the owners and/or passengers of vehicles: e.g., first and last name, telephone number, email address, musical preferences, preferred temperature, and conversations.

2) Technical data relating to the vehicles and their components: electromechanical (e.g., engine temperature, condition of the brakes, functioning of the sensors, angle of the wheels), hardware-related (e.g., use of computers and memory), and software-related (e.g., operational performance, failures, and versions).

3) Geolocation and journey data: e.g., position, speed, destination, route, date, and time.

4) Sensor data (e.g., from cameras and lidar devices) and the processing of this data (e.g., calculated trajectories, obstacles detected, and commands sent to actuators).

178 Federal Ministry of Transport and Digital Infrastructure, "Automated and Connected Driving," Ethics Commission, 2017, www.bmvi.de/SharedDocs/EN/publications/report-ethics-commission.pdf?__blob=publicationFile.

The first three groups are not without their issues—or problems in terms of confidentiality and ethics—but neither are they entirely new. As early as 2014, Ford's Director of Marketing and Sales said: "We know everyone who breaks the law, we know when you're doing it. We have GPS in your car, so we know what you're doing."[179] Regulations or recommendations already exist or are in the process of being established in many countries to handle these issues, as is the case in Europe with the General Data Protection Regulation or in France with the work of the CNIL (the French administrative regulatory body "Commission nationale de l'informatique et des libertés") in regard to connected vehicles.[180]

When it comes to the fourth group, autonomous vehicles will represent a full-scale revolution. Petabytes of data will be collected from cities, the countryside, and their inhabitants like never before. If there are a large number of vehicles in circulation and their data is controlled by a few private players, we will find ourselves in an unprecedented situation due to the magnitude of the ubiquitous capture of data in public spaces, the ability to expand this private data capture to ever larger areas, without creating new infrastructure, and the difficulty of informing non-users of the recording technology about the data that is being collected on them.[181]

This new kind of video surveillance system[182] involves risks relating to the protection of personal data and individual liberties, as the example of Uber's "God View" has shown.[183] This feature made it possible to know everything about a given customer of the ride-sourcing service and to follow their journeys in real time without their permission. It was used by the company's employees to spy on customers of their choosing, including politicians, celebrities, friends, and ex-spouses.

This data capture also raises the issue of the appropriation of the value and

---

179 "We know everyone who breaks the law, we know when you're doing it. We have GPS in your car, so we know what you're doing." quoted in Bruce Schneider, *Data and Goliath: The Hidden Battles to Collect Your Data and Control Your World*, W.W. Norton & Company Inc., New York, 2015. Jim Farey, the Ford executive who made this statement, then backtracked on it, saying "I absolutely left the wrong impression about how Ford operates. We do not track our customers in their cars without their approval or consent."

180 CNIL, *Pack de conformité, véhicules connectés et données personnelles*, 2017.

181 Cara Bloom, Joshua Tan, Javed Ramjohn, and Lujo Bauer, *Carnegie Mellon University, Self-driving cars and data collection: Privacy perceptions of networked autonomous vehicles*, USENIX Association, Thirteenth Symposium on Usable Privacy and Security, 2017.

182 This system actually includes more than just traditional video, since there are other types of sensors in addition to cameras.

183 Will Evans, "Uber said it protects you from spying. Security sources say otherwise," *Reveal*, 2016, www.revealnews.org/article/uber-said-it-protects-you-from-spying-security-sources-say-otherwise.

income generated by the data collected, as well as the issue of managing matters of public interest. Without a strict legal framework, it is probable that this mine of information—which is also a gold mine[184]—will fuel the players that are most advanced in terms of monetization[185], rather than the most legitimate players in political terms, unless strict rules are laid down from the outset. This has already become an issue with applications such as Waze: the data collected in real time is used to provide trip recommendations where the best route for the individual (i.e., the fastest trip for each user), which is created using the criteria of avoiding areas of high traffic, does not necessarily mesh with the wishes of the wider community.[186] It will become even more urgent when most vehicles and other means of transportation are connected and accessible through a single app (see Mobility as a Service in Chapter 9, Systems and Services). At that point, recommendations will be made based on the logic of optimization, with the aim, for example, of maximizing the profit made from the package of services offered. The algorithms will have a similar impact on mobility as road signs, the geometry of the roads, or road regulations, which are currently controlled by public authorities. The way in which the authorities regulate "algorithmic transportation" will be a major challenge in the years to come, and will only become more significant with the arrival of autonomous vehicles.[187]

Financing

The arrival of autonomous vehicles on the market will disrupt the economy and the jobs of the transportation industry, leading to massive needs for retraining and reorientation. By way of example, the relatively high safety of these vehicles should reduce the need for repairs, which will in turn trans-

184 The value of the information collected is much higher than the cost of collecting it, which is almost negligible.
185 McKinsey has already published a document on how to monetize vehicle data: *Monetizing car data. New service business opportunities to create new customer benefits*, 2016.
186 Suburban American cities, such as Leonia, New Jersey, or even Los Altos Hills, California, have begun to close off some of their streets to non-resident traffic in response to the influx of vehicles generated by Waze recommendations. In Israel, where Waze was founded, a suburb of Tel-Aviv is suing the company for creating traffic jams in its region. See John Surico, "What Happens When a City Bans Non-Resident Drivers," Citylab, April 18, 2018, www.citylab.com/transportation/2018/04/the-small-town-that-took-on-waze/558215.
187 This type of regulation exists in the airline industry. See Official Journal of the European Union, Regulation (EC) No. 80/2009 of the European Parliament and of the Council of January 14, 2009, on a Code of Conduct for computerized reservations and repealing Council Regulation (EEC) No. 229/89.

form the insurance sector and deprive hospitals of organs for transplants.

If the arrival of autonomous vehicles goes hand in hand with the electrification of vehicles on the road, as is likely or even to be hoped for, the proceeds from fossil fuel taxes will drop sharply—at a time when investment in road quality is most needed. What will replace this source of revenue? It is not yet known.

At the same time, autonomous vehicles that routinely respect all traffic rules will gradually drain the revenue from speed cameras and parking fines. They will also reduce revenue from city center parking lots, as they find cheaper places to park away from high-density areas.

Conversely, and although this is still up for debate, the use of autonomous vehicles may require the massive deployment of new communication networks (i.e., smart roads to accommodate intelligent vehicles), which will have to be financed.

The depletion of the fuel tax base is an opportunity for smarter and more efficient road financing: tolls reflecting the "real cost" of mobility (e.g., infrastructure financing and external costs of congestion, pollution, and use of space), which will vary according to where and when the vehicle is driven (and parked), and which will be easy to collect (e.g., if the connected vehicle itself is a connected meter that tracks everything). These tolls will vary greatly depending on the region: driving in major city centers, where space is scarce, will cost much more than in the outskirts and rural areas. Mobility pricing is one of the challenges of the autonomous electric vehicle; pricing must be built from the ground up, along with its acceptance.

Public space management

Local authorities are responsible for devising and driving public policy in their regions. It will be up to them, first and foremost, to set up the forums for dialogue with their constituents and to consult with them on the benefits of accepting autonomous vehicles, the conditions for their implementation (e.g. public space law), the (contrasting) consequences of the choices to be made, which depend on whether individual or more collective visions are favored, and the conditions for a transition to a mobility system where car ownership would no longer be necessary.

They will also be responsible for providing their regions with the necessary infrastructure. The use of autonomous vehicles in complex urban networks may require tools for communication between vehicles and infrastructure, either throughout the city or at key locations. Local authorities may or may not choose to maintain control of the digital versions of “physical” road signs, including traffic commands (e.g., stop and no-entry signs). They may also decide to have a hand in the onboard information and entertainment services offered to vehicle occupants.

These local authorities will also have to manage the public space according to new requirements: the need to reserve spaces for recharging electric vehicles will increase with the growth of the fleet; delivery areas will need to be designed to accommodate new logistical practices; and it may be necessary to start reserving curbside areas for pick-up and drop-off of users of on-demand transport. Most importantly, there will be a need for a way of informing users of a public space that driverless vehicles are operating in the area and reminding them of the relevant rules.[188] The main issue here is not, as with traditional public transport, a matter of protecting a bus route from private cars that could slow it down, but rather of avoiding potential conflicts with pedestrians. A symbolic language will need to be invented to allow pedestrians and autonomous vehicles to coexist in the same space, thanks to a shared sense of civility, rather than coercion.

---

188 For example, one manufacturer is proposing to equip autonomous vehicles with "eyes" that are able to signal to pedestrians that they have been “seen” by moving.

Chapter 8

Use

A Historical Disruption in Travel Time Budgets?

In this chapter, we will analyze the impact of the autonomous vehicle on individual choice in terms of mobility habits, based on two optimistic assumptions: 1) the technical and economic promises have been delivered on, and mass-produced Level 4 vehicles (driverless in specific situations) or Level 5 vehicles (driverless in all situations) are available at a similar cost to the user as equivalent traditional vehicles (slightly higher purchase price, but lower running costs, thanks in particular to lower insurance, fuel, and parking costs)[189], and 2) we assume that the various trials conducted in a wide variety of contexts have been successful from a safety perspective. Potential users' misgivings have been allayed, and the vast majority of people are willing to use a self-driving car, in the same way that they are willing to use elevators or automated subway lines today. Consumer trust is no longer an issue.

Under these assumptions, the autonomous vehicle is set to become part of the range of transportation resources available to end users (e.g., those with travel needs), businesses (e.g., to meet the needs of their employees or customers, for deliveries, and for mobile business activities), and the authorities responsible for providing public transportation for all. This integration could take any number of different forms, some of them similar in spirit to the main systems in use today (e.g., self-driving passenger cars, self-driving buses), while others are inspired by more niche current systems (e.g., on-demand transportation by robo-taxi using a smartphone app, or truly individual, single-seater vehicles, like the "pod" developed by Catapult[190]). Eventually,

---

189 Patrick Bösch, Felix Becker, Henrick Becker, and Kay W. Axhausen, *op. cit.*

190 "Self-Driving Pods," *Catapult Transport Systems*, 2018, ts.catapult.org.uk/innovation-centre/cav/cav-projects-at-the-tsc/self-driving-pods.

it will be part of a system that includes other modes of transportation (e.g., subways and railways, pedestrians, and cyclists, but also human-driven cars, at least for a while). On initial assessment, then, there will be a multitude of solutions available to users. How will they choose between these options?

Research shows that there are four main factors behind a person's preferred mode of transportation: 1) cost, 2) time, 3) the travel experience (e.g., comfort[191], ease of use, accessibility, cleanliness, availability, emotional connection to the transportation mode), and 4) social reasons.[192] The last item reflects the fact that individuals transmit non-verbal messages through their choice of transportation. For example, consider what driving a muscle car may suggest about someone's social status, preferences, or style compared to, say, always traveling on foot, by public transport, or on a scooter.

We will begin by focusing on the impact of self-driving cars on motorized transportation. As things currently stand, we have several types of vehicles that offer different types of services. From a usage perspective, we can distinguish between four scenarios (see Figure 6 on opposite page) depending on whether:

- the vehicle is used by a single person or by several people simultaneously (e.g., ride-sharing and carpooling); and
- the vehicle is used by a single person or by several people consecutively (e.g., shared ownership of the vehicle or car sharing, taxi and private hire services).

The arrival of autonomous vehicles will shake up the status quo by altering those four key factors that affect individual choice, which we will now go into one by one.

Cost

The immediate outcome of introducing driverless vehicles will be, perhaps obviously, the absence of a driver. Services that previously relied on drivers

191 Ten minutes of being crammed in with 6 people per square meter may be "worth" 20 minutes of sitting comfortably.
192 Linda Steg, "Car use: lust and must," in Talib Rothengatter and Raphael Denis Huguenin (eds.), *Traffic and Transport Psychology: Theory and Application,* Elsevier, Amsterdam, 2004, pp. 443–452.

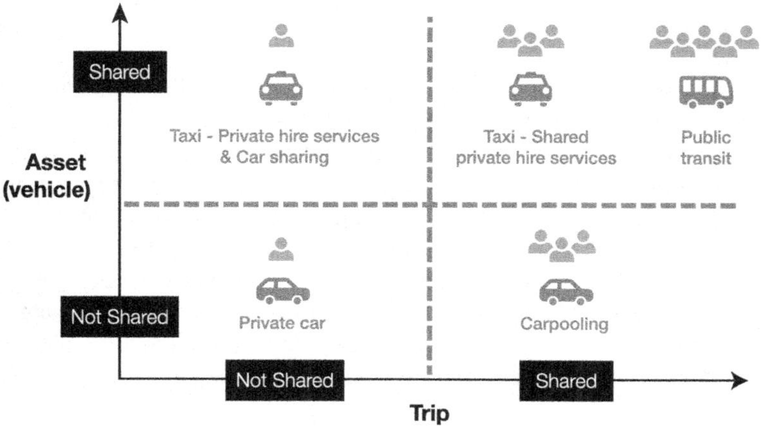

Figure 6. Motorized transportation mode types

Source: Yann Leriche, Jean-Pierre Orfeuil.

(such as taxis, private hire vehicles, and public transportation vehicles) will see their production cost drop, especially since the cost of the driver makes up a significant part of the cost per mile. There will be an even more significant reduction in costs for non-shared modes (i.e., where the user pays 100% of the cost of the driver, and therefore saves as much in the driver's absence) than for those that are shared (i.e., where the cost of the driver is split between all the passengers).

In a driverless context, the difference in the cost per mile between a taxi (or private hire vehicle) and a private passenger car is a result of two types of costs:

- booking costs, fleet management costs (cleaning, lost items, etc.) and customer costs, which are higher for taxis; and
- fixed vehicle costs (spread over fewer miles in the case of individual use) and parking costs, which are higher for private vehicles.

In sum, self-driving taxis are likely to be more cost-effective for people who don't often drive, particularly in parts of the world where the cleanliness of these vehicles is respected (cleaning is a major expense for car-sharing services, and one that is largely overlooked by studies on autonomous vehicles[193]). For frequent drivers, these taxis will be roughly equivalent in cost to private cars.

As soon as they are shared (by several people traveling at the same time, at the cost of a few detours or having to reach the pick-up point on foot), taxis and private hire vehicles will see their cost per mile drop significantly. This is probably the main transformation that motorized transportation will undergo with the arrival of autonomous vehicles. It is highlighted the most frequently, along with safety, when people advocate for the introduction of autonomous vehicles: the possibility that in the future it would be cheaper to travel from door to door (or thereabouts) via taxi than using one's own private vehicle. In this new paradigm, which we will discuss further in the next chapter, cars would mainly be used as a "service" and would no longer be owned.

With regard to cost as a deciding factor when it comes to choosing a method of transportation, two observations should be added. The first relates to the fact that when choosing between a private vehicle and another mode of transportation, the immediate costs of the vehicle (fuel, tolls) are generally taken into account, but the other costs (depreciation, insurance) that have already been paid are not. This favors the use of one's own vehicle, even if it is more expensive when the full costs are taken into consideration. The second observation relates to the future use of the private autonomous vehicle: rather than being just a personal vehicle, it could be used by the whole family or even a circle of close friends. Its ability to travel on its own will enable it to take a friend to work early in the morning, then return home to pick up the children and drive them to school, before taking a retired family member to a gym class and finally picking up food from a grocery store that has a drive-through for autonomous vehicles. This shared use within a circle of trusted people would reduce usage costs, without generating the additional costs associated with public use (e.g., dirt and internal damage).

193 Patrick Bösch, Felix Becker, Henrick Becker, and Kay W. Axhausen, *op. cit.*

Time

As we've seen, autonomous vehicles should lead to an increase in the throughput of infrastructure. This increase could result in a reduction in travel times in several ways, as long as an increase in traffic does not negate the positive effects of automation. First, being freed up from the task of driving should improve the traveling experience and increase the distances covered (see the following point). Second, autonomous vehicles will provide or restore access to motor transportation for those who currently do not have any, such as minors, people with disabilities, the elderly, and those without driver's licenses. And last, the costs of delivery services would be reduced, once they can be provided without a driver, with the resulting effect of increasing usage and traffic.

Given those factors, it is difficult to predict what impact the autonomous vehicle will have on traffic conditions[194], especially when their driving settings have not yet been determined (i.e., will vehicles be configured to always drive at the maximum authorized speed, or will there be thresholds that are specific to each user or operator?). Nevertheless, traffic conditions will have no impact on the future distribution of different modes of transportation, with the exception of vehicles traveling on dedicated lanes (i.e., buses, streetcars, trains, and taxis), which could lose their advantage if traffic in general were to flow more freely.

However, it will no longer be necessary to waste time looking for a parking space or parking the car: the vehicle will take itself to the parking lot, or even continue to drive if it is cheaper to do so, until the next trip. This benefit will be even greater when the trip is not shared. When it is shared, only the driver would save this parking time, because he no longer has to go and park the car (the other occupants of the vehicle can be dropped off at their destination the same as today).

---

194 The conditions could also vary according to the location: more free-flowing on expressways, comparable or more difficult in urban environments.

Travel Experience

Today, most journeys are made in private vehicles with the user in the driver's seat. Traveling is therefore an activity in its own right[195], which "costs" time. While Zahavi's conjecture remains poorly explained, we can well understand the mechanism behind it. We want to spend time at different places throughout the day, but our ability to go wherever we wish runs up against a barrier: the time that must be spent to transport us in a 24-hour daily setting. So a balance is struck between the time spent at a destination and the time spent in mobility. This balance has not changed over time because the use of dominant modes of transportation (e.g., walking, traveling by horse or even by train[196]) has, until now, been an activity in itself.

With the autonomous vehicle, traveling will no longer be an activity that takes up all or most of the time of the person doing it. It could be possible to do many different things in an autonomous vehicle, with the main limit being our imagination: we could read, have a work meeting, write a book about...transportation, play video games, nap, go for a run (in an autonomous vehicle equipped with a treadmill), use a sauna, etc. The ubiquity of digital connectivity makes the possibilities endless. A whole range of activities that were previously only possible in fixed locations will become possible while on the move. The same is true for services: why waste time going to see a primary care physician for a health check-up when all you have to do is order a "doctor vehicle" using an app, which will pick you up at home and examine you before dropping you off at your destination?

Under these circumstances, will the Travel Time Budget (TTB) remain the same? Most likely not, and this will represent a major disruption. TTBs will be able to increase, since mobility and activities will no longer be mutually exclusive concepts as they are today. A long commute to work or a long trip to the doctor or a health center will be easier to accept, because the time it takes will no longer be wasted, or at least less of it will be.

However, TTBs cannot keep increasing forever, as they are subject to

195 Even if some additional activities are still possible: eating a sandwich, making hands-free calls, etc.
196 It is possible to carry out these activities while on a train, but they are limited by the presence of other people and the internal layout of the vehicles. In addition, travel by rail involves extra time in addition to the journey time itself (getting to the station, waiting, etc.), during which the number of activities that can be performed is also limited.

three limitations. Firstly, there are only twenty-four hours in a day, and enough time must be allowed at a given destination to make the journey there worth taking. However, there is some room to grow in terms of the current situation, where TTBs are approximately one hour (i.e., one hour of travel per day). Secondly, the cost of transportation still has a monetary component that will increase with distance. There is likely to be some room for growth here too, especially if autonomy reduces the cost of travel per mile. Finally, mobility is not just a matter of traveling between places. The sequence of activities planned also needs to be taken into account. For example, the time available between a parent finishing work and the time the childcare center closes represents a limit to the amount of time the parent can spend traveling.

How much more time will we spend traveling in the future? It is impossible to say at this point, given the many factors involved (e.g., future traffic conditions, the size of autonomous vehicles). Theoretical studies have illustrated this perfectly, as each of them produces different results. However, the first field trial has confirmed that the potential for growth in mobility with the arrival of autonomous vehicles is significant.[197] Led by a team from the University of California, Berkeley, the study aimed to anticipate how the lives of several households would change if equipped with Level 5 autonomous vehicles by providing them with a driver free of charge for 60 hours a week. The main difference between an autonomous vehicle and a traditional vehicle is that the user does not have to drive or even be in the vehicle when it is moving around (e.g., when it is parking, refueling, picking up pizzas, or recharging). A driver makes it possible to "simulate" this mode of use without waiting for the autonomous technology to reach the necessary level of readiness. The study showed that the distance traveled by these households increased by 83% compared to their previous usage.[198] The number of long trips (over 20 miles) and evening trips after 6 p.m. increased by 91% and 88%, respectively. Retired people stepped up their usage of these types of trips the most, with an increase of 175% and 246%, respectively. Non-passenger trips accounted for 21% of the increase in miles traveled. The answers given most often by participants asked about the reasons for this increase are unsurprising: 1) Productivity:

197 Mustapha Harb, Yu Xiao, Giovanni Circella, Patricia L. Mokhtarian, and Joan Walker, *Projecting Travelers into a World of Self-Driving Vehicles: Estimating Travel Behavior Implications via a Naturalistic Experiment*, Draft, November 15, 2017, presented at the Transportation Research Board 97th Annual Meeting (January 2018).
198 The experiment was carried out on a small sample group. The orders of magnitude obtained are certainly accurate, but the values are not.

the ability to make use of the time spent traveling; 2) Trips without a passenger: the vehicle's ability to park, refuel, or bring home groceries on its own; 3) Convenience: without the need to drive, people are able to make trips in their vehicle more often, even when tired or under the influence of alcohol.

If the machine provides a driving experience as smooth as that offered by a human driver, and doesn't cause motion sickness[199], autonomous vehicles will have a positive effect on the travel experience, by freeing up time, allowing travel when incapacitated or not in possession of a permit—and their ease of use (e.g., it will come to pick you up wherever you are and can refuel by itself). This positive impact will be enhanced by the fact that the time freed up by not having to drive can be spent doing other things. This will be the case if the layout of the vehicle and arrangement of the passengers in it allow each person to do their chosen activity. Accordingly, other activities will be easier when traveling alone in a self-driving car that is fitted out to meet the passenger's specific needs, rather than in a bus. It is therefore the individual use of the autonomous vehicle that will offer the most benefits in terms of time used for other activities; therefore individual use will very likely become more attractive than uses that involve sharing with third parties.

Social Motivations

Our choices are not just a matter of mobility (see the box below). They are also guided by our egos and a desire to convey a certain image of ourselves or to adhere to social norms. These social motivations, coupled with emotional motivations (e.g., our emotional relationship to the vehicle), significantly contribute to explaining the variations in car use.[200]

What image and values will private autonomous vehicles convey in the future? What about the private self-driving taxi, the shared self-driving taxi, or the self-driving bus? At the moment, it is too early to say. One thing is for sure: the way in which they are introduced, in particular the tone of

199 This subject is covered in: Mohamed Elbanhawi, Milan Simic, and Reza Jazar, "In the Passenger Seat: Investigating Ride Comfort Measures in Autonomous Cars," *IEEE Intelligent Transportation Systems Magazine*, Vol. 7, No. 3, 2015.
200 Linda Steg, *op. cit.*

the rationale behind it, will be a determining factor in the future use of the autonomous vehicle.

The Culture of Mobility, the Mobility Experience, and the Travel Experience

People choose what mode of transport to use based on the criteria of instrumental reasoning (e.g., travel time, cost of travel). However, this is not the whole picture.

Our decisions are also embedded in the collective culture of mobility. In the 1960s, using a car (and nothing else) was a sign of middle-class status. Then came cars that were marketed directly toward women, a sign of women's liberation in the 1970s. Today, distancing oneself from the personally-owned car and joining the shared vehicle movement is trendy for the "creative class"[201] and among some millennials, but the cultural and emotional attachment to a personal and privately-owned car remains strong among the suburban middle class.

These choices also depend on each person's biographical trajectory. A person's mobility experience leads him or her to select certain travel means as their set of possible choices. Someone who has never used public transit will not be aware of the schedules and whether the system can meet their needs, and therefore will usually eliminate it as a potential option.

The travel experience introduces qualitative elements (e.g., perceived reliability, physical and psychological comfort) into the decision-making process. A key factor in this experience is whether the trip or vehicle is being shared or not. With the current dominant mode of travel (the private car), nothing is shared; we are in a private world, one that's closer to home. With the traditional alternative mode

201 Here, we are using the archetypes proposed by Thibault Muzergues, in *La quadrature des classes* (Le bord de l'eau, Bordeaux, 2018).

(public transport), everything is shared, within a familiar environment: public space, where civil inattention is the norm. Carpooling, which requires a small space to be shared by two or more people, is still only used by a minority of the population, and has developed its own set of rules (e.g., the option to take trips where passengers chat with one another or make journeys that simply get you from A to B). The sharing of vehicles is also practiced by a minority. The space is private during the journey but traces of previous users may be left behind (e.g., perfume, smoke, and litter).

The transport mode promoted by the proponents of the autonomous vehicle (i.e., shared robo-taxi services) entails a double adherence: the sharing of the vehicle and the sharing of the trip in a space offering more intimacy and privacy than public transport, without the presence of a person enforcing certain rules, such as a bus or taxi driver. In terms of the travel experience, this is more similar to being in an elevator than on public transport, but an elevator ride will rarely last more than a minute. Time will tell if a new form of sociability will emerge in these environments, where passengers are kept in close proximity to strangers for longer periods of time.

Summary

From an individual perspective, two possible futures are emerging.

1) The first is one where people no longer own vehicles because autonomous transportation services are practical and cost-effective, thanks to the sharing system. The advent of this era is aptly described by Christian Fritz:

You have just bought your first fully autonomous car and let it drive you to work this morning. It was a little scary at first, but you forgot about the lack of a driver quickly, and now you are at your office. What is the car doing right now? Right now, it is sitting in the parking lot, doing nothing other than perhaps charging. What a waste! So you decide to let your car act as a cab when you are not using it. You don't want to run a cab business, you just want to reduce the cost of car ownership, so you outsource the job to a cab service. Only, you're not the only one thinking like this. Everyone's got their car moonlighting as a cab. There is an over-supply, and no one's making much money. In fact, there are so many cheap cabs, you sell your car and just use autonomous cabs to get around instead. It's much cheaper than owning your own.[202]

This first vision of the future assumes that the shared service vehicle will be the dominant mode of transport. This is a big assumption, which cannot be taken for granted, even at a time when the sharing economy is growing (see the box on following spread).

2) The second vision of the future is one that involves privately owned autonomous vehicles, which would mostly be used without ride-sharing, just as they are today. For most individuals in developed countries, the time component is the most important determining factor in choosing a mode of transport.[203] The value assigned to time[204] increases with the person's standard of living. Given that time cannot be extended, people are increasingly trying to optimize it. Individual ownership and use of an autonomous vehicle would effectively transform the time spent traveling, which was previously wasted time, into useful time. If the currently high price of autonomous technologies does fall as expected, then the extra cost of owning an autonomous vehicle compared to using an autonomous service vehicle would be offset by the value of the time that would be made useful. To this list we can add an unbeatable ease of use: it will be practical (there is no need to park it or to avoid drinking alcohol), it will always be available[205] (it can follow the user whenever he or she uses a different mode of transport, for example, to carry his or her belongings), and it will be personalized (outfitted

202 Christian Fritz, "Mobility-as-a-Service: Turning Transportation into a Software Industry," *Venturebeat*, 2013, venturebeat.com/2014/12/13/mobility-as-a-service-turning-transportation-into-a-software-industry.
203 And for the wealthier classes in developing countries.
204 This is the amount we are willing to pay to save a specific amount of time, for example, an hour. For the same individual, this value can change at different times of the day or year, depending on the person's activities and the constraints involved in linking those activities together one after the other.
205 In dense urban areas, service vehicles will arrive shortly after they have been ordered; they will probably take longer to arrive in more sparsely populated areas.

according to the owner's work or leisure needs).

Of these two visions of the future (which will probably coexist), which one will end up dominating? Among the most credible assumptions[206], there is a scenario where the autonomous vehicle will start out primarily as a service vehicle, because it is both expensive (the technology will take time to become affordable, car sharing reduces user costs) and limited in its functions (it could be restricted to a geographical area of operation at first, before that area is extended by technological improvements), and it will be owned later on, when costs have fallen sufficiently and its performance has improved.

Beyond individual preferences, the uses of autonomous vehicles, and also of other transport modes (e.g., public transit, walking, and biking), will be shaped by the systemic functioning of mobility in the future.[207] This systemic functionality is what we will explore in the next chapter.

206 See B. Gordon, S. Kaplan, F. El Zarwi, J. Walker, and D. Zilberman, *The Future of Autonomous Vehicles: Lessons from the Literature on Technology Adoption*, University of California, Berkeley, 2018.
207 This systematic operation depends on the stakeholders involved: the individuals we have already mentioned, but also local authorities, manufacturers and mobility service operators.

Does the Age of Access Mean the End of the Ownership of Consumer Durables?

This is Jeremy Rifkin's theory in his landmark book: "For a long time, capitalism has been identified with ownership. The market is the place where goods are exchanged. The explosion of telecommunications and the new economy have led to an unprecedented transformation: markets are giving way to networks, sellers and buyers are being replaced by suppliers and users. The increased use of licenses, leases and subscriptions sounds the death knell of ownership."[208] The theory is justified by the rapid obsolescence of products, as well as the ease of

208 Jeremy Rifkin, *The Age of Access*, G. P. Putnam's Sons, New York, 2000.

access provided by the Internet. Another book, What's Mine Is Yours, which was published ten years later, had a similar impact, placing a strong emphasis on B2B markets and "service products."[209,210] Consumers are in turn riding the wave of access, via platforms, sharing, and peer-to-peer collaboration, boosted by the increased opportunities for communication offered by the smartphone. A few chapter titles ("Enough Is Enough," "From Generation Me to Generation We," and "Community Is the Brand") reflect this instrumental and axiological trend.

These theories are undoubtedly supported by the facts: the rise of streaming, the replacement of ownership with access to digital goods (e.g., recorded music), and the growth of all kinds of platforms (Airbnb, BlaBlaCar, Drivy, Uber, Autolib), to name just a few examples from everyday life. However, does this also mean the virtualization of consumption, the victory of access over ownership, of collaboration over private use and attachment to consumer durables? This is doubtful. The replacement of tens of thousands of public phone booths with personal cell phones could refute this idea, as could the proliferation of screens that allow each person to connect to their favorite media at home and to take their digital self with them whenever they go out. Despite the rise in house prices and the relatively weak financial situation of many in the younger generations, home ownership has continued to grow in France (although at a slower pace). The number of houses with swimming pools has almost tripled since 2000 (2 million compared to 700,000). The residential square footage per person has increased by a third in twenty years, and the number of second homes has continued to grow (3.3 million compared to 2.6 million twenty years ago).

In the U.S. home ownership has remained quite stable, at 63.9% in 1985 and 64.2% in 2018. However, the square footage of living space per person in a new house has increased 92% since 1973,

---

209 Rachel Botsman, Roo Rodgers, *What's Mine is Yours. The Rise of Collaborative Consumption*, Harper Business, New York, 2010.
210 For example, replacing the purchase of a photocopier with a rental model that comes with a maintenance service, paper and ink supply, etc.

from 551 square feet to 1,058 square feet.[211]

Uber customers in the Île-de-France region (greater Paris) own more cars than average Île-de-France residents.[212] Uber users also tend to have a higher income.

French households owned 27 million cars in 2000, which rose to 34 million in 2015. This 7 million increase is in stark contrast with the fewer than 100,000 cars that are available for rental between individuals and for car-sharing purposes. In the U.S., registered vehicles increased by nearly 38 million in the same time period, from 225.8 million in 2000 to 263.6 million in 2015.

Increasingly, cars are being used by solo drivers, not only in France, but also in the United States, the birthplace of carpooling. In the U.S., working at home is the fastest growing sector (in percentage terms). It's also this mode that involves the least sharing of experiences in the workplace, at the water cooler, in interactions between coworkers. Clearly, in all of these areas, car ownership has more than stood its ground.

These findings bring three observations to mind. The first is methodological: we must bear in mind the scale involved, which has only a marginal correlation with the trends highlighted by the media and social networks. The second is more operational: the move toward less ownership and more sharing is certainly something to be wished for, but the idea that it will happen naturally through generational change is simply wishful thinking, which can only lead to disillusionment. The same is true of the development of platforms and the arrival of new technologies, such as the autonomous vehicle. The third is more theoretical: if, as François Ascher said, owning and renting is like having your cake and eating it too,

211 The American Enterprise Institute, a public policy think tank dedicated to defending human dignity, expanding human potential, and building a freer and safer world. Blog Post by Mark J. Perry, Scholar, June 5, 2016. https://www.aei.org/carpe-diem/new-us-homes-today-are-1000-square-feet-larger-than-in-1973-and-living-space-per-person-has-nearly-doubled/
212 6t-bureau de recherche, *op. cit.*

it is because the user experiences are different, as is the ability of each to meet one's situational needs. There is clearly much room for improvement in our understanding of this area.

	1980	*1990*	*2000*	*2010*
Drive Alone	64	73	76	77
Carpool	20	13	12	10
Motorcycle	6	4	3	3
Public Transportation	6	5	5	5
Work at Home	2	3	3	4
Other	2	2	1	1

Table 6. Commuting modes in the United States (%)

Source: Commuting in America 2013, AASHTO, 2015

Chapter 9

Systems and Services

Sharing or Owning: a Key Issue

A threefold selection process will determine the services provided by the autonomous vehicle when production is up and running: 1) the benefits to users (in terms of cost, time, travel experience, and social benefits, as we have seen), but also 2) the pressure applied by public and private players to organize and promote its development, and 3) the legitimacy of the justifications given by these players, as perceived by the public.

Clearly the concrete benefits to passengers will become apparent, particularly the ability to use one's time in a variety of more productive and enjoyable ways given that one will no longer have to focus as fully on driving. That said, in a context where expert opinions could vary significantly, we can assume that the legitimacy of the justifications provided by the various players will not be achieved by simply bombarding the public with information, however plausible the justification may be (e.g., the autonomous vehicle will drastically reduce accidents), but by choosing new opportunities for AV development where clear benefits that far exceed the costs or risks can immediately be felt. These benefits may relate to the image of local municipalities or regions and their elected representatives (see the BCG study that makes the case for disruptions in the Paris transportation system ahead of the 2024 Olympic Games[213]), or they may relate to real improvements in individual mobility.

213 Megatrends that can be found in the Boston Consulting Group's partnership with players in the Île-de-France region, which has nevertheless taken the precaution of asking questions and exploring the necessary institutional changes: "Mobility Nation. Comment l'Île-de-France peut-elle réinventer sa mobilité d'ici à 2030. 24 propositions pour Paris 2024," *bcg.com*, November 6, 2017, www.bcg.com/fr-fr/d/press/6november-france-press-release-175580.

The need for legitimate rationales that consumers accept will likely have a major consequence: the trajectories of AV development are likely to differ greatly in different parts of the world. We can therefore assume that the personal-use autonomous vehicle will be more likely to prevail in countries with steep social inequalities that are assumed as legitimate and that, by contrast, shared use will dominate in countries where solo driving is already subject to heavy restrictions. Furthermore, in the current context of globalization, it must be anticipated that the success or failure of AV trial rollouts in one part of the world will have an impact on elected representatives in other parts of the world, and that these results will be leveraged by supporting and opposing sides in ongoing debates.

Similarly, the pressure applied by public and private players to promote and coordinate the development of autonomous vehicles will vary from country to country, depending on the strengths and weaknesses of the existing transportation systems in each geographical region.

Due to these factors, we will begin this chapter by briefly outlining the current state of mobility in countries like the U.S. and France, before going on to explore what services the autonomous vehicle might offer in response.

Car-Dominated Mobility and the Desire to Escape it

Not surprisingly, the car's continued dominance is still the major factor in the U.S. According to the Pew Research Center, 88% of all U.S. households own a car[214] (France is not far behind with 83%), and in 2017 there were an average of 1.88 vehicles per U.S. household.[215] A recent survey by the American Automobile Association Foundation for Traffic Safety showed that the average U.S. driver spends 51 minutes a day behind the wheel, 20 min-

214 Study by Pew Research Center surveying car ownership in 44 countries. Italy leads with 89% of households owning cars and the U.S. follows with 88%. https://www.citylab.com/transportation/2015/04/global-car-motorcycle-and-bike-ownership-in-1-infographic/390777/

215 Summary of Travel Trends: 2017 National Household Travel Survey. U.S Department of Transportation, Federal Highway Administration. https://nhts.ornl.gov/assets/2017_nhts_summary_travel_trends.pdf. Cars per household on Table 2a on Page 9.

utes more per week than in 2014, and drives some 31 miles per day.[216]

In terms of the 5.7 trillion person miles traveled in the U.S in 2016, 70% of these miles occurred in cars or other personal vehicles, 24% by plane, and 5% in public transit.[217]

A handful of U.S. cities are investing significantly in public transit and significantly improving their systems. However in many other U.S. cities, there is insufficient frequency and reliability in transit services, which continues to force many people to either lease or buy a car, or find a ride with a friend or family member, and to suffer through traffic congestion.

French mobility surveys offer a good indication of the established modes of transport.[218] Public transportation only plays a major role within the Île-de-France region, which includes Paris (21% of people use public transportation in the overall Île-de-France region, while 34% use transit in Paris—three times higher than car usage in Paris).

In terms of the percentage of total trips in the remainder of France, only 6% of trips are made by public transit, and the majority of the population does not use it. The share of trips by public transport peaks at 10% in the heart of large urban areas and falls to 3–4% in rural or suburban towns. It is therefore advisable to avoid extending the lines of argument that are relevant in the context of Paris to the country as a whole. Walking still accounts for 22% of local journeys, but it only represents 2% of the distances traveled.

Regarding new, emerging modes of transportation, surveys are beginning to provide an idea of their significance. According to the ObSoCo Institute, 3% of the population in France used car sharing in 2016, 5% used peer-to-peer car rentals, 5% used a bicycle sharing system, 10% used scooters, skateboards, or roller blades, etc., 10% used driven transportation services, and 30% used carpooling over the course of the year.[219]

---

216 American Driving Survey conducted by the non-profit American Automobile Association Foundation for Traffic Safety, surveying nearly 12,000 drivers between 2014 and 2017. https://newsroom.aaa.com/2019/02/think-youre-in-your-car-more-youre-right-americans-spend-70-billion-hours-behind-the-wheel/

217 Transportation Statistics Annual Report 2018 published by the U.S. Department of Transportation, Bureau of Transportation Statistics. https://www.bts.dot.gov/sites/bts.dot.gov/files/docs/browse-statistical-products-and-data/transportation-statistics-annual-reports/Preliminary-TSAR-Full-2018-a.pdf Page 3-1 and 3-2.

218 They describe the mobility behavior of the population aged 6 years and over during a working day.

219 We are using information from ObSoCo (a Paris-based consumer research body focused on emerging consumption): Observatoire des mobilités émergentes, vague 2 (Emerging Mobility Study, Phase 2), September 2016.

Car-Related Problems

The problems experienced with cars vary greatly from one region or metropolitan area to the next. In large cities, a major problem is that the automobile can detract from the quality of daily life. People, especially those from city centers, want a city that is calmer and less polluted and that also flows better with fewer traffic jams. In the suburbs and outlying areas of these cities, public transportation services are often considered to be quite poor, so individuals are heavy car users, but they declare themselves to be "motorists through force of circumstance."

In outlying towns, suburbs, and rural areas, the combination of a high dependency on cars[220] and frequently (but not always) long travel distances often means that the car takes up a significant proportion of the household budget, especially in low-income households. Even if this share is partially or wholly offset by lower housing costs, the absence of public transportation options for all is a source of resentment.

The strong car regulation initiatives that have been implemented in a few cities (including Paris, London, and New York) are interpreted as restricting commuter access to downtown and the city center.

Beyond these problems, there is a fairly widespread feeling in France and Europe that "we cannot go on like this"—that we must move away from the dominance of the car or the dependence on it—because of the frustration related to highly congested traffic, the prospect of climate change, and the objectives of transitioning toward renewable sources of energy that many people support.

While it is not our intention here to debate this issue, we must nevertheless consider this aspiration and its immediate implications. There are two frames of reference for guiding public action: one being the problems experienced by current users, and the other the energy transition, which advocates for a reduction in personal car use. Although these frames of reference may converge on certain points, there are also significant areas of friction or even contradiction.

---

220 It is estimated that about 40% of the French population is not served by public transportation.

Incorporating the Autonomous Vehicle into Regional Policies

Local authorities are expected to manage their region or metropolitan area in the best way possible to benefit the lives of their residents. They must advance local initiatives that will make their region attractive and successful at bringing in new jobs and attracting qualified employees, as well as tourists, visitors, and retirees, all of whose spending contributes to the local economy. Many city planners believe that the pursuit of these objectives must be done in accordance with a goal to transition to a sustainable, less carbon-generating economy.

A region's appeal depends on multiple factors, the importance of which varies from place to place, but transportation issues are always high on the agenda. A strong public transit system is often a very important element in employers' decisions about which city to choose for a location or to expand their operations.

In addition to its mobility infrastructure, cities and regions must be "visible" (i.e., they must appear on the mental map of stakeholders, which often implies regional marketing efforts), easily accessible (hence elected officials' interest in being close to highways and having access to rail or air services), and should have the necessary infrastructure (hence the interest, for example, in region-wide broadband communications infrastructure).

The objectives in terms of internal mobility within the regional area are also largely shared, but the efforts and political will needed to ensure successful regional approaches to mobility can differ. Key mobility goals include:

- Providing a high level of accessibility for all to essential resources, employment, education, healthcare, shopping, and leisure. For larger cities, providing this access is rarely a problem. For others (including the suburbs of major cities), distributing these essential resources, including servicing them by public transportation, providing virtual access (i.e., tele-services), and offering mobile services (such as mobile libraries or mobile health care checks) can contribute to the solution.

- Providing mobility solutions for all, at a cost that is compatible with the resources available to local authorities.
- Promoting solutions that have the least negative impact on air quality in the region and the smallest possible carbon footprint.
- Promoting mobility solutions that take up less space and support the development of peaceful places to live in high-density areas.
- Participating in road safety initiatives.
- Contributing to public health objectives by promoting physical activity.

These goals are ambitious, and the road to get there is anything but straightforward. Elected officials, who are in office for finite terms, must commit to long-term strategies that address complex situations with many targets to be met within a limited budget.

To What Extent will the Autonomous Vehicle Help in Achieving these Objectives?

In situations where the autonomous car is owned, it will contribute, in its own way, to the objective of mobility for all, e.g., by facilitating travel for teenagers and people who might have been reluctant to get their own license. It will help to improve road safety. It will have a positive or negative effect on air quality and initiatives to minimize carbon footprints depending on whether or not it is electric.

The owned AV will not help reduce traffic or the use of space (since we will drive more) or contribute to public health objectives (i.e., the car will become more attractive, so there will be even less of an incentive to walk or cycle[221]).

Even when the most promising scenarios for the autonomous vehicle are taken as a given, its impacts on society will not all be positive. In the current context, however, there is no regulatory tool that would prevent an individual from acquiring an autonomous vehicle on their own account, especially since insurance companies have repeatedly stated that they are prepared

221 However, better traffic safety could have a positive impact on "soft" modes of transportation, i.e., sustainable and active means such as walking or cycling.

to assume civil liability. If the AV does fulfill its promise of offering a safe, self-driving vehicle at no or little extra cost, then it is destined to become widespread—with major consequences for suburban expansion (we will explore this issue further in the next chapter).

Local authorities are therefore more likely to try to either deter or ignore rather than encourage the use of private autonomous vehicles, at least in terms of solo use.

The objective of controlling the use of space can only be achieved by ride-sharing, thanks to its better use of transportation infrastructure. However, this optimization will not automatically lead to a reduction in road traffic. If traffic can initially be reduced thanks to a better vehicle occupancy rate and therefore improved traffic flow, new demand will be generated.

This induced demand is the direct consequence of Zahavi's law: we seek to maximize the opportunities offered to us, not our time budgets. If traffic flows more smoothly, some people will take their cars for trips they would not otherwise have made or would have made using a different mode of transport (e.g., a more sustainable mode such as public transport), thereby adding to traffic and congestion.

Trips could be shared either through carpooling or through public robo-taxis, shuttles, buses, or heavy-duty modes of public transportation. However, as discussed in the previous chapter, we do not believe that the experience of traveling in a shared autonomous vehicle is comparable to the experience of traveling in a private car. It would seem wrong, then, to suggest that if the time spent traveling is comparable[222], and even if the journey is cheaper when shared, it would not matter to users whether they were in their own vehicle or a shared autonomous vehicle. The presence of others and the lack of personalized features will hamper one's ability to transform travel time into useful time.[223]

222 A strong hypothesis, given that ride-sharing with third parties generally requires detours, or a less direct route than the one that would have been used alone, in order to accommodate the various passengers' destinations.
223 These discussions are nothing new. In 2010, Sharon Shewmake, in an analysis of the decline in the percentage of commuters using carpooling in the United States, despite the construction of 2,300 miles of HOV (high-occupancy vehicle) lanes reserved for vehicles carrying more than one person, wrote: "One reason for the decrease in carpooling rates, is that carpooling comes with an additional set of personal costs. These costs include reduced route flexibility, grouping costs, and a loss of privacy when another person shares the car. Encouraging carpooling may not improve traffic conditions as much as advocates claim, since new carpoolers may be people who would otherwise not have driven." She then goes a step further and wonders: "Are HOV lanes truly a traffic control measure, or are they a way of expanding capacity without violating the Clean Air Act Amendments?" Sharon Shewmake, *Can Carpooling Clean the Air? The Economics of HOV Lanes, Hybrid Cars and the Clean Air Act*, University of California, Davis, MPRA paper, No. 26129, 2010.

For shared use, it will therefore be necessary, on the one hand, to understand how to reduce the discrepancy in travel experience (for example, by developing attractively designed vehicle interiors) and, on the other hand, to offer incentives for the shared use of vehicles or, conversely, to introduce penalties for empty or individual use. From a technical point of view, these schemes could be applied easily to connected autonomous vehicles if based on financial factors. The vehicles will be able to transmit information about the number of passengers it is carrying (e.g., zero, one, or more), which will be used as a basis for taxation and for payment purposes. Politically speaking, shared AVs will have a harder time gaining acceptance if there are penalties involved, especially during the transitional phase when traditional and autonomous vehicles coexist. Taxation of a vehicle at different rates based on its occupancy rate is currently very rare[224], tricky to impose on existing roads, and complex to implement on a large scale for non-connected vehicles.

The battle for shared use by coercion is therefore difficult to win. Would it be easier to use incentives? If the incentive is financial, it is likely to only have a limited impact for the reasons given above. This effect has been observed in situations where an urban public transportation network has switched to free travel: there is only a slight change in modal share. A time-based incentive is more promising, since time is the key factor in choosing a mode of transportation for many people. This could take the form of dedicated lanes for vehicles carrying several people or reserved lanes for public transportation. This reserved space is the fundamental principle behind mass transit, which would remain unchanged by autonomous vehicles: the most efficient way to transport a large number of people in a restricted space is to transport them together in large vehicles on dedicated lanes.

A Cure to "Cost Disease" in Public Transport?

What will change, however, are the economic conditions under which these public transportation services can be operated. When urban shuttles, buses,

224 This practice exists on some U.S. highways, where the fastest lane is reserved for paid use, the cost of which decreases according to the number of people on board. A special monitor would have to be installed in the vehicle to track the number of passengers. Gantries placed above the roadway would detect this information, which would then be used for billing purposes.

and streetcars are made autonomous, they will be freed from U.S. economist William Baumol's "cost disease" which they currently suffer from. This effect applies to services that experience little or no growth in productivity. In his first studies, Baumol focused on the performing arts. Their productivity gains are zero: a staging of Macbeth lasted 2 hours and 15 minutes in Shakespeare's time, and the same is true today. Public transportation[225] has the same characteristics: operating a regular bus service from point A to point B invariably requires a driver, a bus (a much less significant factor in the economic equation than the driver), and fuel. Any special adjustments to accelerate the speed aside, it always takes the same time to complete the journey, and there is very little room for any gain in productivity. In order to avoid a situation where employees move to different industries with productivity gains and rising wages, "stagnant" sectors must increase wages at the same rate, but without benefiting from any productivity gains that would reduce or cancel out the impact on the cost per unit of the products or services sold.

Thus the cost of public transportation is inexorably increasing, unlike the cost of other products or services, such as cars. To keep prices low, for social and urban efficiency purposes, local authorities have offset this increase by constantly increasing their subsidies (see Figure 7). For a long time, these services were private (like theaters) and for-profit, before they gradually came under public control to keep them in operation as their financial difficulties increased.

The scale of the discrepancy between investment and usage can be measured by comparing the levels of use of systems against the public expenditure allocated to them. In France, in round figures, in 2015, $31 billion was spent on public transportation and $15 billion on roads. This means that roughly two-thirds of total spending went to public transit, which delivered some 4% of total trips. In the U.S., in 2018, Americans took 9.9 billion trips on public transportation. A total of $70 billion was spent on public transportation in 2017, and $177 billion was spent on roads. So in the U.S., 28% of the total spending of $247 billion was invested in public transit, which delivered about 5% of total trips. These figures show that while our roads in both countries would be severely more choked without public transportation, it does require significant investment.

225 And more generally, all services whose production relies on manual labor. Before William Baumol, Jean Fourastié had already shown this in his statistical series. Jean Fourastié, *Productivité, prix et salaires*, Organisation Européene de Coopération Economique, Paris, 1957.

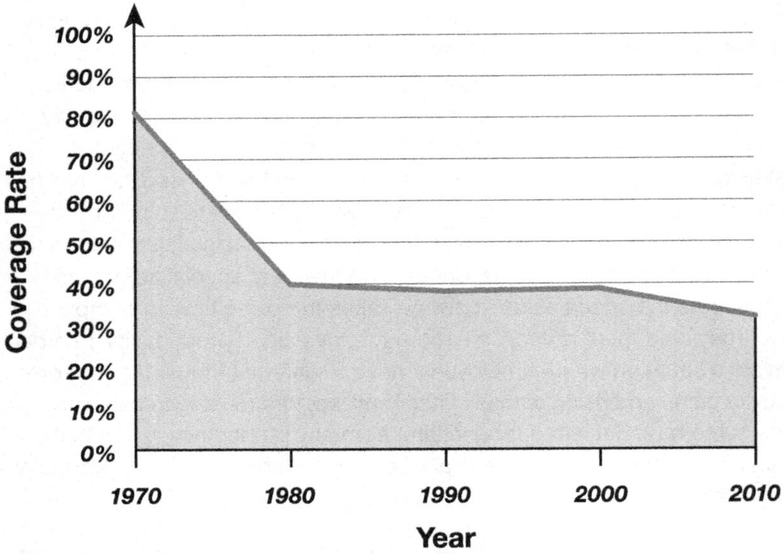

Figure 7. Change in farebox recovery ratio (coverage of the operating expenses by fare revenue) of public transport networks in the U.S. from 1970 to 2010

Source: American Public Transportation Association Fact Book, 2019 (ATPA Fact Book 2019) Appendix A, Tables 87 and 68.

In this context, autonomous vehicles present a unique opportunity for local authorities: a public transportation system that no longer suffers from the "disease" of rising costs. But what types of services would this cover?

Beyond mass transit mentioned above—which, for geometric reasons, is and will remain the most efficient way to transport a large number of people in high-density areas—public transport also has a social purpose: to provide access to low-cost motorized transportation for many people. Mainly ensured by regular bus lines today, this mission could be carried out tomorrow by autonomous vehicles (of different sizes), operated on fixed lines or on demand. Today, transit authorities manage these services, either directly or by contracting the services of private partners. It will likely remain that way in the future, if AVs are predominantly individually owned and used: local authorities will continue to supplement private modes of transportation with public services that are accessible across the region.

In the scenario where autonomous vehicles become predominantly shared, service-oriented, and managed by fleet-owners, local authorities will no longer need to be directly involved in the supply (at least with regard to public transport with a social role to play). Abundant and available to all, public transport will not need to be complemented and improved. Public authorities will maintain only a regulatory role—to ensure the equity of the transport supply available in their region (i.e., the coverage and availability of private transport services) and its accessibility (financial, physical, and cognitive)—which they will exercise through contractual and financial mechanisms yet to be devised. For example, the authorization given to a robo-taxi operator in a region may be accompanied by a service quality obligation for any point in that region, and not only in the dense areas, particularly regarding the maximum time spent waiting for a vehicle.

Shared Mobility at Regional Scale

Is the scenario where mobility is predominantly ensured by robo-taxi or robo-bus fleets, offering individual or shared trips, at the scale of a city of agglomeration, technically credible? The answer is yes. Several academic studies have shown this in different contexts.[226] One of the most ambitious studies was conducted by the Organisation for Economic Co-operation and Development (OECD).[227] Carried out in the city of Lisbon, Portugal, it started with the region's real transport demand and modeled several mobility scenarios. In the case where the subway is retained (but not the bus lines), it showed that all of the transport demand could be served door-to-door by a fleet of autonomous taxis for shared trips with two, five, and

226 Lawrence D. Burns, William C. Jordan, and Bonnie A. Scarborough, *Transforming Personal Mobility*, Earth Island Institute, Columbia University, New York, 2013; Jaison Zachariah, Jingkang Gao, Alain Kornhauser, and Talal Mufti, *Uncongested Mobility for All: A Proposal for an Area-Wide Autonomous Taxi System in New Jersey*, Princeton University, Princeton (NJ), 2013; Daniel Fagnant and Kara M. Kockelman, "The travel and environmental implications of shared autonomous vehicles, using agent-based model scenarios," *Transportation Research Part C: Emerging Technologies*, No. 40, March 2014, pp. 1–13; Paolo Santi, Giovanni Resta, Michael Szell, Stanislav Sobolevsky, Steven H. Strogatz, and Carlo Ratti, "Quantifying the benefits of vehicle pooling with shareability networks," *Proceedings of the National Academy of Sciences of the United States of America (PNAS)*, Vol. 111, No. 37, September 2014; Kevin Spieser, Kyle Treleaven, Rick Zhang Emilio Frazzoli, Daniel Morton, and Marco Pavone, "Toward a systemic approach to the design and evaluation of automated mobility-on-demand systems: A case study in Singapore," in Gereon Meyer and Sven Beiker (eds.), *Road Vehicle Automation*, (Lecture Notes in Mobility), Spring 2014.

227 OECD, International Transport Forum, *Urban Mobility System Upgrade: How shared self-driving cars could change city traffic*, Corporate Partnership Board Report, 2015.

eight seats, managed by a central dispatch system. The number of AVs required was ten times less than the number of cars currently present in the city. The number of kilometers traveled by the AV fleet was somewhat higher than that of by all current human-driven vehicles (buses and cars): +9% at rush hour and + 6% on average throughout the day. Traffic therefore increased in relation to the existing situation, due to the disappearance of buses and the appearance of trips by passenger-free vehicles. The need for parking spaces experienced a spectacular drop: the projected need was only 5.6% of the existing availability, enabling an urban space equivalent of 170 soccer fields to be reallocated for other activities.

Other models were constructed by the OECD after this first study. Their work from 2016, also conducted with data from Lisbon, produced even more spectacular results because it was carried out with larger vehicles and reduced limits concerning trip times.[228],[229] This study concluded that "congestion disappeared, traffic emissions were reduced by one third, and 95% less space was required for public parking in our model city served by Shared Taxis and Taxi-Buses."

These models are nothing more than exercises in fluid mechanics, and they cannot claim in any way to reflect a future reality. The study methods assumed a fixed mobility demand, which does not grow with supply, thereby ignoring the phenomenon of traffic induction. However, these studies are useful because they show that shared AV services are able to meet the demand for door-to-door transport without increasing individual travel times in an unacceptable way due to excessive waiting times or detours. For example, in its first study, the OECD limited additional transport times and distances to 20% above those of the direct route, with ceilings of 10 minutes (including the time spent waiting for the vehicle) and 2 kilometers, respectively.

The credibility of a future where mobility is mainly ensured by shared vehicles is reinforced by the emergence of a second revolution, alongside that of autonomy, in the world of mobility: that of "MaaS," or mobility as a

---

228 However, the OECD studies are based on a strong hypothesis: all vehicles are "dispatched" from a central station, which makes it possible to optimize how the journeys are grouped. On-demand shared transportation systems benefit from strong network effects in the regions where they operate—the higher the density of supply and demand, the more efficient they are. In the real world, however, there is competition between operators, which will undoubtedly lead to the formation of oligopolies.

229 OECD, International Transport Forum, *Shared Mobility, Innovation for Liveable Cities*, Corporate Partnership Board Report, 2016.

service. MaaS concerns developing services that enable access to several transport modes (e.g., chauffeured car, taxi, rental vehicle, public transport, bicycle, or scooter) in a simple way, via a unified digital interface.

The individual passenger car is the dominant transport mode today because its strengths are unparalleled. With it, we can travel basically all distances, in all directions, alone or in a group, most often door-to-door, and without having to make any transfers to other modes of transportation. In most cases, it is the fastest mode, with the exception of the motorcycle.[230] The car's driving can be adapted according to the circumstances, from very relaxed to more nervous driving when time is short. If necessary, baggage, purchases, and tools can all be left in the vehicle when it is parked. And owning a car makes it possible to provide support services to others.

Car ownership and use have a cost that is accessible to the greatest number of people in the developed world. Being inside a car offers a high level of protection compared to other road users, such as pedestrians and cyclists, and it protects users from bad weather during the journey. It is a cocoon that maintains privacy. The owner can adjust the seats according to their morphology and choose the desired atmosphere (e.g., temperature and music). They are the master of the degree of internal cleanliness of the vehicle (which they generally ensure themselves) along with its bodywork. It enjoys a dynamic and well organized used car market. A user can get rid of a car when they wish and replace it if they have the means to do so, without losing too much money (which is not the case with other durable goods, such as sofas or televisions). In these circumstances, it is hard to find an alternative mode of transport to the individual car that is able to challenge its hegemony.

However, a group of several easily accessible alternative modes could upset the established order. Or at least that is what MaaS promises: the ability to plan a multimodal journey, book it, pay for it, and make the whole thing so simple that owning a car—or a second car—is no longer indispensable for households. Fleets of shared autonomous vehicles will benefit from the development of MaaS platforms. These platforms will present their options alongside other modes, which together will be more attractive than each one of them independently when compared to the individual car.

---

230 However, the rise in 30 kilometer per hour zones reduces the time and competitiveness gap for trips that do not involve expressways.

A third element could give credence and promote the advent of mobility mainly ensured by robo-taxis: the emergence of new business models for the operators of these fleets that would reduce the cost of mobility services and therefore increase their attractiveness. There is no reason to assume that the mileage or hourly price of future AV transport services can be calculated by taking the direct cost of these services and adding the operator's margin. The digital economy is full of models that do not respect this rationale. Some services are even free; their suppliers (e.g., Facebook, the Google search engine, and Twitter) finance themselves via the data collected on their customers, data which may not in fact reflect their customers but their products, as the saying "if you're not paying for it, you are the product" suggests.

It is easy to imagine that comparable business models would be developed by fleet operators.

1) Because vast amounts of data will be collected, AVs will act as veritable crawlers (or spiderbots) of the real world. They will provide their operators with information that could be of interest to local authorities (e.g., data about traffic flows, the condition of the roads and infrastructure, video surveillance without having to install fixed cameras, and detecting vehicle violations without having to use a traffic officer) and providing private companies data about the potential volume of customers in a given geographical area for potential sales, targeted advertising, map surveys, and so on.

2) However, mobility is a "derived demand," as we have seen. It is the cost associated with accessibility to what we care about that is the primary demand. If the destination we are planning to reach has an interest in us making the trip, it may finance some or all of its cost. This rationale is already in place in some city centers, where shop owners pay for their customers' parking tickets. Tomorrow, IKEA, Walmart, Whole Foods, and others could make the trips to and from their stores free of charge by entering into an agreement with the AV fleets servicing them.

Autonomous Vehicles: for Better or for Worse?

MaaS (alongside, in all likelihood, new business models for operating robo-taxi fleets) will promote service-oriented and shared mobility. Will that be enough to resolve the issue raised in the previous chapter and to confirm the AV's position as the dominant model of mobility for the future?

We do not think so. The mobility ecosystem is made up of not only individuals, local authorities, and vehicle fleet operators, but also car manufacturers and all kinds of other motor vehicle manufacturers. Autonomous technologies and their subsequent uses will change the design and functionality of the objects that will circulate in our streets. The autonomous vehicle is not a conventional vehicle without a driver. When owned individually, it could be extensively personalized, and it could better meet the needs of its owners than the personal vehicle of today. The manufacturing sector for vehicles and other wheeled objects will deploy its creativity and resources to respond to all specific requests, from the office-robomobile to the bedroom-robomobile through the follower-robomobile, whose aim will not be so much to transport its owner but rather their personal belongings or purchases.[231] Therefore, mobility via privately-owned autonomous vehicles, with a predominantly individual use, has considerable chances of prevailing once the cost of the technology declines.

The question of the use of autonomous vehicles once production is up and running is also a question of the future of our cities. Robin Chase[232] perfectly expressed this with her statement: "Self-driving vehicles will improve our cities. If they don't ruin them."[233] AVs have the potential to increase road flows, to transform our living environments by freeing up practically all parking spaces in roadways, to enable the greatest number of people to go from point A to point B quickly and safely for the price of a bus ticket, and to improve air quality and significantly reduce carbon emissions. But if the only change that occurs is the elimination of the driver, the results might not be as positive: the distances traveled will lengthen; congestion

231 The Gita "mobile-carrier," a luggage-hauling robot made by Piaggo, is one of these designs: www.piaggiofastforward.com/gita.

232 Robin Chase is the co-founder and CEO of Zipcar and author of the book *Peers Inc. How People and Platforms are Inventing the Collaborative Economy and Reinventing Capitalism*, PublicAffairs, New York, 2015.

233 Robin Chase, 2016, *op. cit.*

will beat all records, reinforced by delivery vehicles and "zombie" passenger-free cars; resources from communities will dry up with the loss of taxes on fossil fuels, parking revenues, and parking fines; and unemployment will increase with the loss of employment by professional drivers.

These are serious issues, and some of them (particularly that of employment) go beyond the scope of this book. While the technical aspects of taxation remain to be explored, it is easy to imagine the solutions that could be implemented by leveraging the connectivity of the vehicles. The same cannot be said when it comes to traffic and congestion issues. As we have described, shared robo-taxi fleets, often cited as the best future solution, will only have the effect of increasing the human throughput of road infrastructure, but will not guarantee a reduction in traffic and congestion. Traffic conditions influence urban transformation processes, which in turn affect the demand for mobility.

Finally, it should be noted that there are many other possible futures besides those that are publicly envisaged. To take just one example, we could plausibly imagine lanes reserved for Level 4 autonomous vehicles, along the lines of the HOV/HOT[234] systems in place on some U.S. highways. The principle would be as follows: the driver leaves home behind the wheel of their car. Upon arriving at the highway, control of their vehicle would be taken up by a system ensuring an autonomous circulation in the form of a platoon. This platoon-like circulation would produce speed gains, a dynamic toll would enable for the costs of the system to be covered, and it would guarantee the flow of the lane considered. The user of the service would enjoy both significant time savings and an opportunity to occupy their time with something other than driving. The driver would then take back control of the car when leaving the highway to reach their final destination. This form of toll has the advantage of not being imposed on everyone, and the disadvantage of making condition-related inequalities visible.[235] In this scenario, which recalls the concept of "first-class roads" studied in the 1980s[236], automakers would sell just as many cars,

234 HOV/HOT: High-occupancy vehicle/high-occupancy toll lanes: a system of reserved lanes on highways for high-occupancy vehicles (buses or cars occupied by at least two or at least three people) and open to solo drivers paying a toll.

235 Public transit is further along in this plan to differentiate services according to users' ability to pay, as shown by the decision to create a premium service, CDG Express (a planned railway line to Charles de Gaulle airport in Paris), as a complement to the existing regular line.

236 See Francis Papon, *Les "routes de première classe": une tarification différenciée de la circulation en agglomération pour en améliorer l'efficacité économique de manière socialement équitable*, Doctoral Thesis (under the supervision of Rémy Prudhomme), Université Paris 12, 1991.

motorway operators would expand their activity, and the (happy) owners of an AV would be able to “forget” traffic jams and retain the pleasure of driving when traffic is flowing.

Chapter 10

Territories / Regions

The Best Scale at which to Manage the AV's Arrival

If autonomous vehicles do live up to their promises, they will help lower the costs of mobility, mainly in terms of time used for individual modes of transport, but also in monetary terms for services that currently require a driver. Overall, then, we can easily anticipate the impact they will have on regions, in continuation of the developments outlined in the first part of this book—namely, increased opportunities for urban expansion and greater distances between people and social groups.

That said, these impacts will not happen as automatically as we might think. First, the relationship between the expansion and structure of space and the performance of transportation systems is not a direct one. Greater mobility has merely enabled the processes of urban transformation observed thus far to take place; it has only been a permissive factor. The driving forces behind it stem from the regional stakeholders: households looking to increase their living space; suburban municipalities attempting to control their populations using urban planning programs that limit new construction, which in turn shifts construction to more distant towns; large-scale retailers relying on raid networks to develop mass offerings adapted to extended targeted sales areas; employers able to move away from city centers; peripheral regions experiencing agricultural decline developing new ambitions; and so on.

The first issue at hand therefore centers around the future development of these driving forces.

The second impact concerns the nature of future mobility itself. The autonomous vehicle will blur the line between mobility and immobility (or sedentariness) when it becomes possible to carry out a wider variety of tasks while on the road. A historical look serves as a reminder that we must be humble when it comes to predicting the urban transformations induced by innovations in mobility (see the box below).

Our aim here is not to make forecasts or provide answers to these issues, but to reason from the fulcrum that seems to us the most robust: vehicles will be used in a context where they provide some benefit, that is, where they facilitate mobility.[237]

And that approach is why we have chosen an analytical framework that differentiates between regional levels, since the challenges involved are different at each level. Following in the logical footsteps of Marc Wiel, we distinguish three main categories of mobility: local, urban agglomeration, and supra-metropolitan mobility.[238]

---

237 With the understanding that their contributions will be subject to the places and modes of operation (private use, shared taxis, autonomous shuttles, etc.) that have been authorized.
238 Marc Wiel, 2007, *op. cit.*

Innovation in Mobility:
From the Urban Transformations Expected to those that Occurred when these New Technologies Were Adopted, by way of a few Examples

The elevator: a technology that has not reached its full potential and has led to unexpected transformations

Moving vertically has always been the most tiring for humans, so much so that very early in history, people tried to devise ways to move loads vertically. This mode of transport experienced an outstanding development with coal mines. However, it was only in the mid-nineteenth century, after the demonstration of a safety feature by Otis in New York, that elevators became an integral part of urban daily life.[239] At that time, cities continued to grow in height and to become more dense. One could then have expected that this technology would have allowed the upward momentum to continue, particularly since technical advances were enabling the construction of increasingly higher buildings. This opportunity was fully exploited for service activities, especially in business districts,[240] where residential buildings were few. As a result, elevators significantly contributed to the functional specialization of urban areas. But the elevator remained under-exploited in the domain of housing, due to the development of individual housing. Its presence nevertheless helped to maintain occupancy in collective buildings, in the long periods when individual residences were in vogue, and to homogenize the way people occupied buildings. Before the elevator, high floors were reserved for service staff and poor people, while wealthier citizens resided in the lower floors.

In summary, the elevator has helped to maintain the number of people living in apartment buildings and has contributed to the functional and social segregation of space.

Public transportation: from a corporate strategy catering for market needs to a strategy of urban de-densification driven by public authorities

Leaving aside the unfortunate experience of Blaise Pascal and his five-sol coaches (five-coin coaches)—along with the first public transportation services in Nantes that would transport customers from the center of the city to a heated bathhouse—regular public transit services first appeared in France in Paris' business district. They were initially created and developed on a purely private and competitive basis. However, the

239 From the point of view of the mobility experience, the elevator is an interesting mode of transportation because it has points in common with shared autonomous vehicles: it's autonomous, it arrives on demand, and the trip can be shared with others (who greet you when you enter, which does not happen on the subway). It is by far the most widely used means of public transportation in France (about 100 million trips per day), with an incident rate (people getting stuck) of five incidents per million trips. Fatal accidents are rare, but highly publicized.

240 Which, with their skyline, can sometimes become the symbol of a city.

public authorities were soon called upon to regulate competition, and they used these services to further their own aims. Their main objective was to make every effort to reduce density in the center of the capital, beginning by granting operating permits in areas of high-density demand, on the condition that less attractive neighboring districts would be served, with the goal of making it easier for people to live in outlying districts. Faced with the difficulties of negotiating with each company, they created a monopoly, an omnibus company, which also created a de-facto network effect and had an impact on the system as a whole, paving the way for technological advances, from horse-drawn transportation to the subway by way of streetcars. In the last quarter of the nineteenth century, local authorities went on to introduce a "workers' tariff," a basic requirement that would allow workers of modest means to leave the areas nearest their jobs and move to the outskirts of the city, or even its immediate suburbs. As a result, public transit became a tool for urban planning.

The sidewalk and curb: from regulations governing use to forms of usage that shape regulations

On the eve of the Revolution of 1789, the sidewalk barely existed in French cities. In the century that followed, an increase in vehicular traffic and faster vehicles led to a huge rise in the occurrence of violent clashes between pedestrians and drivers, as well as often fatal accidents. The first half of the nineteenth century saw priority given to speed, the separation of pedestrians and vehicles via the general proliferation of sidewalks, and the obligation for pedestrians to use them and not encroach on the roadways. At the time, this obligation was a hazardous task; people walking next to buildings were exposed to rainwater from roofs and often residents' wastewater.

The sidewalk is no footnote in the history of infrastructure, not only because of the function it fulfills (providing access to buildings), but also because of its footprint. In Paris, for example, sidewalks take up 40% of the road surface, with the remaining 60% being used by vehicles of all types. However, this was not always the case. During the golden age of the automobile, sidewalks were largely overrun by illegally parked cars, which rarely received tickets. As public opinion has turned against the use of cars in cities, the simple introduction of metal posts prohibiting cars in certain areas has been a major factor in the decline in traffic in Paris since the beginning of the millennium.

The curb also plays an essential role in urban life. From today's perspective, it is easy to imagine how sidewalks could have been bordered by bike lanes. However, bicycles were not very widely used when the sidewalks were created, and there was even less appreciation of the safety issues affecting them. When bicycle use began to spread, the space had already been taken. The curb of the sidewalk was theoretically used as a stop where vehicles could drop off their passengers or goods: a French law of 1851 and a decree of 1921 prohibited "vehicles from parking needlessly on the public road."
The law did not prevent long-term parking on roads from spreading (especially close to people's homes), thanks to the great leniency shown by the public authorities.[241] An order by the Paris Prefect of Police in 1948 (followed by an act in 1954 that extended the provision to include the whole country) enshrined the supremacy of custom over formal law, removing the obligation to keep vehicles' headlights and taillights on at night on public roads.

This leniency lasted as long as the size of the fleet did not exceed the parking space available. That limit was quickly exceeded, and shop keepers, who were supporters of the "no parking, no business" slogan, asked the public authorities to crack down on long-term parked cars. In France, the Land Use Act then allowed local authorities to require new buildings (e.g., residential and office buildings) to build parking lots,[242] and various regulations ("blue zones," paid parking, priority spaces for residents) made it possible to better manage parking on public roads. In

---

241 This feeling of ownership by citizens of the public space around their housing is reminiscent of the situation that prevailed until the 18th century, when "the street seemed to belong more to the inhabitant than to the passerby" and gives an idea of what may become of the "NIMBY" (Not In My Back Yard) phenomenon. It reminds us that even in metropolitan areas, highly localized oversight plays an important role. Thus, when lanes are too narrow to accommodate two-way parking and two lanes of traffic at the same time, it is generally preferable to make all the lanes go in one direction, rather than remove one lane.
242 This will make it possible to boost car ownership and car use for commuting to work.

doing so, parking policies broke a taboo ("Why should we have to pay for parking when our taxes have already paid for the roads?") and introduced an accepted form of discrimination (where residents are given priority over other people).

If it seems like these policies still lack coherence, it is because they have conflicting objectives: while people may want fewer cars locally, they also want to keep businesses in their region that involve the use of cars to some extent, particularly when these businesses face fierce competition from other businesses, such as out-of-town shopping malls.

All in all, these unclear and contradictory policies, alongside other developments[243], have helped maintain the vitality of city centers, in a context where the automobile could have brought about a winner-takes-all scenario by adapting all environments to suit its needs.

243 Such as the development of tourism in cities.

Regions and Autonomous Vehicles

Local Mobility

Local mobility is mobility that travels to all the most commonly used services replicated in many places within the city or the metropolis. These are the shortest trips we make (in terms of time and distance); we choose the easiest, closest options, since we have a choice. When the city is slow[244] and dense (the two generally go hand in hand: a slow and sparsely populated city does not grow, because it does not benefit from the positive effects of co-presence), local mobility involves short distances, for example, visiting a local store or business. When the city is fast and less dense, local mobility no longer necessarily involves short distances, because urban stakeholders (e.g., retailers) rely on easy access to develop more concen-

244 Here, "slow" is shorthand for a mode of mobility that is costly in terms of time, but also potentially costly in other aspects.

trated offerings and to establish a looser mesh over the region.

In a European-style vision, where neighborhood relations can be established without the need of a car, short distances and sustainable modes of mobility are encouraged. This has especially been true since the renewal of cycling, and the emergence of light and electrified travel modes adapted to distances of a few kilometers, have made it possible to envisage a sense of proximity without confining people to a given neighborhood.

In dense urban areas, the aim will be to bring door-to-door speeds of sustainable modes of transport closer to that of cars (autonomous or otherwise) and to reduce the use of the latter. The curb will be more easily allocated to cycle lanes (which could also be used by scooters and other light electric vehicles) instead of being used to drop off passengers or for deliveries to private households[245], which will be limited by road use regulations specific to these activities. Efforts will be made to limit the use of autonomous vehicles in this market to those who are unable to use active modes for reasons of limited mobility. Widespread, easy-to-use autonomous vehicles (in terms of traffic speeds, cost per mile, comfort, etc.), whether private or service (even shared trips), would progressively loosen the mesh of local activities and services, access to which would end up requiring the use of a motor vehicle.

In lower-density areas, autonomous robo-taxi or shuttle services may play a more important role, both in servicing these areas internally and in providing feeder services to major public transportation hubs. However, the spatial extension of the services will be limited to the geographic area of a few municipalities which, thus united, may consider they have succeeded in grouping the primary set of urban amenities and services that are needed by their constituents.

In all cases, it will be necessary to accommodate the circulation of autonomous vehicles in transit throughout the region. They may only be allowed on certain roads and under certain conditions, because the confinement of pedestrians to dedicated areas (once an objective of "slab urbanism") will not be tolerated, and local forms of mobility will not be allowed to encroach too much on major roads reserved for other types of mobility.

---

245 On the other hand, deliveries to local businesses must be facilitated.

Urban Agglomeration Mobility

Urban agglomeration mobility refers to trips made by people with only one possible destination—at the time the decision to take the trip is made—within the labor market they have access to. Therefore, this includes trips for commuting to work, for going to an educational institution, for meeting relatives or friends, or more rarely, for doing some exceptional shopping or accessing an unusual service.

At this scale, mobility plays an important economic role: greater mobility increases the number of jobs that can be easily accessed, and therefore makes it possible to better match the supply and demand of skills. If mobility is too easy though, it leads to urban sprawl beyond the city boundaries, without increasing the number of accessible jobs. Within urban areas themselves, easy mobility also tends to facilitate residential segregation, which can be mitigated to some degree by social housing policies.

The pursuit of reducing the overall cost of mobility via autonomous vehicles will have positive effects in economic terms. Longer commutes to work may be acceptable because they will be less of a waste of time, creating opportunities for workers to find jobs that are better suited to their skills.

The consequences of increased traffic on congestion once we have autonomous vehicles remain an open question. AVs' reduced reaction time and their ability to coordinate themselves will increase the flow of roads—under uncertain conditions—as mentioned in Chapter 6, "Technology," provided that they are not excessively mixed with conventional traffic.

During the transitional phase, if journeys get longer as a result of reduced mobility costs without an increase in throughput, massive gridlock will only be prevented by a shared transportation system with large vehicles in dense areas (in dedicated lanes for greater efficiency on high-demand corridors) and smaller vehicles in peripheral areas. The system will have to operate throughout the day in order to convince those who can to give up their personal cars.

In the longer term, a balance will need to be struck between daily and residential mobility: even in large cities, the appeal of expanding labor market areas will reach its limit when commuting distances become too long and too much to handle. At that point, it makes more sense to facilitate residential mobility. This balance will only be achieved through a combination of transportation policies (taking advantage of the opportunities offered by vehicle connectivity), urban planning (providing suitable housing near major employment centers), and taxation (avoiding penalizing residential mobility with real estate transfer taxes that make no economic sense).

Supra-metropolitan mobility

This refers to mobility that goes beyond the traditional labor market area. It connects different conurbations (extended urban areas, consisting of several towns, suburbs, etc.) by providing a link to further amenities and services, but the most strategic element relates to intercompany connections.

This type of mobility is crucial for businesses that are part of the global economy, i.e. companies whose suppliers, partners, and customers are not exclusively based in the local area. Long distance mobility makes it possible to develop strong local expertise and create international centers of excellence. Trips may involve motor vehicles as well as other modes of transport, such as road and rail, but also air travel.

Autonomous vehicles can play an important role in facilitating and developing this form of mobility. In an increasingly fast-paced economy, the speed of transportation is key for numerous trips. Self-driving vehicles (personal or high-frequency minibuses) will reach their full potential if they travel in dedicated lanes (where they may be subject to specific tolls), which will also enable them to travel faster than human-driven vehicles and perform various other activities without compromising on safety levels. Since highway driving does not pose as much of a technical challenge as urban driving, prior to the arrival of Level 5 autonomous vehicles, we may well see Level 4 vehicles that are set to autonomous mode while on inter-urban expressways and switch back to manual driving mode while in cities. There will be demand in both the transportation and logistics sector. For the latter in particular, the risk of inducing significant additional traffic (thanks to an increase in the number of small deliveries, since there will no longer be any driver costs) is very high.

From analysis to action

The impact of autonomous vehicles on regions must therefore be addressed on a differentiated basis (i.e., at several mobility levels, each with different challenges) and in a multidisciplinary way (involving transportation, urban planning, and taxation). The simplistic reasoning mentioned at the beginning of the book, which heaps a whole series of virtues upon the autonomous vehicle, ranging from the elimination of congestion to a massive reduction in energy consumption, can only lead to disillusionment and misunderstanding. Such an approach would prevent us from taking advantage of the benefits that the autonomous vehicle could bring, if it were introduced in a well-thought-out way, taking into account the lessons learned from the past as well as the new possibilities that could be opened up (e.g., innovative taxation through connectivity).

The TRUST analytical framework provides a structure for that well-thought-out approach, taking all of its aspects into account: what are the opportunities offered by technological progress ("Technology")? What practical applications can we expect from it, depending on what the various regulations laid down will permit ("Regulations")? What will be the benefits for individuals; how will they take advantage of the new opportunities; and how will it be used ("Use")? What will the consequences be for the mobility system as a whole ("Systems and Services")? What will be the impact on different regions, and what will be the impact on mobility and our lives due to the rebound effect ("Territories/Regions")?

This means using territories as the starting point (to consider in what kind of regions do we wish to live) in order to understand which kind of mobility is required for their development, and then to define, as concerns the topic of this book, the role the AV should play in this ecosystem, as well as the rules and technologies that will allow it.

The fact that only moderate results have been achieved in twenty years of combating urban sprawl shows that this is not a straightforward task. However, that does not mean the link between land use and mobility is broken. Road and public transportation management have an impact on the forms of regional development. Businesses that involve a lot of freight transportation (e.g., furniture and clothing wholesalers) tend to leave central business districts for the suburbs, due to the difficulties faced by large trucks in dense urban areas, while conversely leisure and tourism activities represent better conditions for urban growth. Employers prefer to hire locally when they perceive commuting conditions as tiring or unreliable. Each municipality wants to ease traffic congestion in its region to provide an environment favorable for the growth of stores and services for its residents. Admittedly, there's never been so much conversation around the city as a concept, but today the city is being built largely through the management of existing infrastructures, and at a very local level.

Conclusion

A General Consensus

It's with a certain sense of caution that we enter the final stretch of this book, given the rapidly changing situation with regard to the autonomous vehicle. When we began our analysis more than 18 months ago, there was a general consensus, based on the announcements made by tech giants, tech platforms, car manufacturers, and consulting firms, predicting the imminent arrival of fully autonomous (Level 5) vehicles on the market.

Since then, statements made by senior auto industry executives have shattered that consensus, and it is now generally accepted that it will be a long time before it will be possible to navigate the Place de la Concorde in Paris in our own personal robot car. However, there is reason to assume that it will be possible to delegate the task of driving under certain circumstances (e.g., on a highway during rush hour) or to launch driverless transportation services on scheduled routes or on an on-demand basis, when the environment allows it, and where the necessary infrastructure is in place to communicate with the vehicles in service.

The prospect of situation-dependent autonomy is clearly no longer limited to Europe or some automakers reputed to be cautious after the statements made by Waymo's CEO, John Krafcik, at a Wall Street Journal conference in November 2018.[246] He simultaneously announced the launch of a commercial self-driving taxi service in Phoenix early in 2019 and his conviction that "autonomy always will have some constraints" before saying that he does not think the industry will ever achieve the highest driving rating, i.e., being able to drive at any time of year in any weather and any condition.

246 Wall Street Journal, "WSJ Tech D. Live: Are We There Yet? The Future of Driverless Cars," November 14, 2018, www.wsj.com/podcasts/wsj-the-future-of-everything/wsj-tech-dlive-are-we-there-yet-the-future-of-driverless-cars/a0c1b34e-fe21-4c19-8b98-8e74b0c0643f.

Although this is a time for caution, investment in autonomous vehicles has continued to grow[247], technology is advancing, and real-life trials are flourishing. The hype surrounding the autonomous vehicle has generated enthusiasm among investors, politicians, and the media. None of the competitors can afford to drop out of the race now, if only because Level 1 and 2 (or even Level 3) vehicles are within reach and don't require any legislative changes on the part of public authorities. It would be self-sabotage for manufacturers to put themselves in a situation where they are unable to offer these innovations to consumers in the near future. Key industrial competitiveness issues are leading countries with domestic manufacturers to put their full weight behind the development of the AV, without getting bogged down in discussions about what the changes brought about by the autonomous vehicle will mean for people and their regions.

Why Should we Plan for the Arrival of Autonomous Vehicles?

Since the autonomous vehicle burst onto the media scene, the situation has evolved, although it is true that the course remains unchanged: we are headed for full Level-5 autonomy, one day! But regional stakeholders now have confirmation that states have priorities other than the AV's impact on regional projects. This situation confers on them a great responsibility. What is truly new and also quite unprecedented for this mobility revolution is that these regional stakeholders can for once escape the tyranny of urgency. They have a longer period than previously announced (but not infinitely extendable) to plan and organize the arrival of this newcomer in our lives as mobile individuals and in our cities, which are already saturated with various types of traffic.

Plan, you say? It may seem a surprising choice of words, particularly when the publicity surrounding the AV up to this point has turned the autonomous vehicle into a matter for the private sector, where manufacturers are fighting it out to create a new product that major consumers will choose to buy. But

247 Cameron F. Kerry and Jack Karsten, "Gauging investment in self-driving cars," *Brookings*, 2017, www.brookings.edu/research/gauging-investment-in-self-driving-cars.

planning is a term we fully embrace, and for a number of reasons.

First and foremost, two matters of principle:

> **1)** Road networks are shared public property, used by a multitude of users with a wide range of needs (from pedestrians to drivers, not to mention street cafés, street performers, and homeless people). What is more, in cities at least, this public property is constantly under pressure: there is no limit to the demands that can be made on the use of this space, but the supply of public space is limited and can be slow to change.
>
> That pressure has increased with the growth of e-commerce deliveries and the emergence of personal mobility devices, such as self-service bicycles and scooters. It is therefore perfectly natural for an arbitration body to decide not only on the rights and duties of each category of users, but also on the incentives (e.g., economic instruments or other tools) that will make it possible to keep demand at a level that can be met by the available supply, under the right conditions. The best candidates for steering this arbitration body are the elected representatives of the regions affected.
>
> **2)** Connected vehicles, whether personal or integrated into fleets of service vehicles, will generate unprecedented amounts of data, not only about the personal lives of the vehicle users, but also about the environments in which they operate. This data can be used by those who collect it, receive it, or transmit it. This is an economic issue, as well as one of citizenship and sovereignty. The fact that citizens continue to broadcast their lives on social networks will not prevent them from turning against public authorities if they feel those authorities have breached their information duty, regarding data flows in normal operations, or their duty of vigilance with regard to potentially illegal operations.

Added to these two principles are a series of reasons related to uncertainties about the promises that have been made: on the one hand, the real benefits of autonomous vehicles have been overestimated, but on the other, its transformative potential has been underestimated. There is no doubt that the autonomous vehicle has qualities that a traditional, human-driven vehicle lacks: it can be used by people who don't have a driver's license, which is good as it helps people exercise their right to mobility. The car doesn't drink alcohol, it doesn't get distracted, and it has better reflexes than humans, so it's good for safety. It can safely lock onto the vehicle in front of it, which helps to reduce traffic jams, so it's good for congestion. If integrated into a transportation service, the AV can drastically reduce the cost of travel (e.g., no driver, no parking to pay for, a vehicle cost that can be depreciated faster thanks to a daily use ten times longer than today, and a trip cost that can be shared with more than one person), so it is good for the wallet. And finally, the AV opens up new ways to spend time while traveling, since a vehicle can become an office, a recreational space, or a studio, etc.

All of these beneficial assertions are partly true, but they are only half-truths, so they are also partly false. We can present counter arguments to each them. Our collective ability to demonstrate the technology's superiority over humans in terms of safety, in all circumstances and based on trials alone, is up for debate. It could be shown that some aspects are too often overlooked at present, such as the travel experience, which could hinder the success of shared robo-taxi systems. It could also be observed that the promoted benefits will mainly be applicable when the entire fleet of vehicles in use has become both autonomous and shared. So, to realize the vision championed by several major consulting firms and tech gurus (i.e., a fleet of robot cars as opposed to car ownership), we would have to accept a transition where one generation would see its mobility capacities degraded, without benefiting from the advantages that would materialize only later on, when the next generation would benefit from these advantages that are potentially important but still being discussed. In the current climate, this timeframe seems unrealistic, and few elected representatives would be likely to accept it.

Finally, the idea that the general level of mobility of the population would remain unchanged if the cost of mobility fell is a pipe dream: history has shown us that mobility increases when its cost decreases. As such, in the absence of any collective oversight, we must expect a continuing increase in distances traveled and a rise in urban sprawl.

The final reason explains the need to oversee and manage the arrival of the autonomous vehicle: players within the self-driving sector and regional stakeholders will have to work together in a coordinated manner. It is possible to change the world without coordinating anyone else when you carve out your own space, as the Internet players have done. It is possible to make a dramatic entrance, like the cowboys of the Wild West, into the on-demand transportation industry in the real world (as opposed to the virtual one), as Uber has managed to do. But the resistance that this company has faced around the world shows that it is only possible to establish a lasting presence in the urban mobility landscape by forging partnerships and sharing objectives with established "old world" players. This is what all players in the on-demand transportation market in the United States are doing today, by diversifying their mobility solutions (from scooters to on-demand car sharing and bike sharing) and by pushing for their mobility solutions to be integrated into regional transportation policies.

The Local Level, as Seen Through the TRUST framework

If we try to examine the major steps that still need to be taken on the journey toward autonomous vehicles (an agenda punctuated by areas of caution) through the lens of the TRUST framework, we can immediately grasp the key role that local and central authorities could play in creating autonomous mobility.

> **1)** In the field of technology, while manufacturers continue to make investments in the hope of winning the global race, states will have to develop a certification process, first for Level 4 vehicles (autonomous under certain circumstances). In this context, states would benefit from developing a typology of regions, conditions, and scenarios by combining deductive approaches (based on regional characteristics) and inductive approaches (based on the situations encountered by the vehicles in circulation) to precisely define use cases. This shared typology does not exist today.
>
> Authorities will also have to decide on the appropriate level of approvals that could be divided among high-level authorities (e.g., a global forum, such as the United Nations Economic Commission for Europe or the European Union, which would probably best meet manufacturers' needs), for issues such as cyber threats, and more

local bodies, which are more aligned with the public's expectations with regard to regional compatibility issues and matters concerning local networks.

From now on, local authorities must therefore be prepared (and prepare their citizens) for their regions to be used as testing sites. They should think about the minimum requirements that will need to be established for those wanting to experiment with autonomous vehicles in their jurisdictions, in terms of data sharing and reporting requirements. With a view to maintaining a living and informed democracy, regional players must start thinking about and creating tools to simulate the consequences of different choices, which will serve as a basis for building shared frameworks and supporting citizen participation.

2) With regard to regulations, local authorities, just like the states, are concerned today with the increasing presence of electric and connected vehicles and the development of delivery and ride-hailing services. In the short term, they will be concerned with trials of autonomous Level 4 fleets, and perhaps in the medium term with fleets of Level 5 vehicles, if they are approved.

The questions that autonomous mobility raises tor regions are numerous:

- How to plan for the integration of these new vehicles and services on the road: e.g., the location of charging stations; reserved spaces for deliveries and passenger pickup and drop-off; and the curbside to be shared by various personal mobility devices. A plan will also need to be developed for repurposing parking lots, once the number of personal cars has begun to decline.
- How the rights and duties of the various users of public space, from pedestrians and scooters to autonomous vehicles, will change. These rules will have to be incorporated into formal codes, but also and more importantly, into behaviors, which once again highlights the need for urban residents to be closely involved in reviewing all the available options and in decision-making processes.

- How to determine pricing and how to finance mobility. The introduction of electric autonomous vehicles will impact revenue from fuel taxes, parking, and fines. State and local authorities must prepare for this financial upheaval, using two keys for understanding the issue: their resource requirements and the role of pricing in guiding urban mobility, both in terms of their levels and the mobility modes used. The widespread adoption of connected vehicles will make it possible to consider highly personalized user charges that reflect the contribution of each user to collective costs and external traffic costs, the main one involving the use of that scarce and non-extendible asset: the road network itself.
- How to use the mass of data that is and will be gathered via connected vehicles, and subsequently autonomous vehicles, by businesses, technology companies, car manufacturers, and service operators.

3) Even if the new forms of usage provide a focal point for autonomous mobility, we can reasonably rule out personal use of Level 4 or Level 5 autonomous vehicles during the next ten years. On the other hand, this decade is an apt timeframe for testing drivers' appetite for a whole series of innovations made possible by the autonomous vehicle:

- Driver and parking assistance features and the impact of their use on mobility.
- The pace of the transition from a private car-based model to a model structured around shared mobility services, and the emergence of new forms of social behavior in connection with these new types of use.
- The pace of development of mobility, focused on bicycles and personal transporters, which will make it more challenging for autonomous vehicles to operate.
- Feedback and concerns from users and pedestrians regarding autonomous shuttles once they are in operation, and how people will cope with the potential for function creep.
- Advances in logistics, which, with the use of delivery robots, could be made more quickly than advances in passenger transportation.

Over the next few years, we will also see new expectations emerge along with the prospect of autonomous vehicles and shared mobility, from those of businesses wanting to attract new customers to the expectations of existing mobile businesses and services, which may foreshadow new ways of occupying vehicles.

4) The issue of systems and services concerns all stakeholders, but local and regional authorities have a particular interest in the promised benefits in terms of road safety, the integration of autonomous vehicles into shared fleets of vehicles (asset sharing), and the shared use of these services (ride sharing), which would significantly reduce the need for private vehicles and parking, as well as congestion related to excessive traffic at certain times of the day.

The capacity of on-demand transportation systems to effectively operate as a shared mobility mode will still need to be tested. In what kinds of urban areas and on what types of routes will the systems be able to combine trips? What level of pressure will be acceptable and necessary in order to encourage a shift from the personal use of private vehicles to the shared use of mobility services? What economic models will make these services viable? Finally, there will also be the issue of the development of delivery services that use autonomous vehicles.

5) Last on the list is the impact on regions, which clearly has a very direct bearing on local stakeholders and will become the primary issue when it comes to taking action. To bring about a collective idea of the region in which we would like to live, at a time when the future is becoming more uncertain and unstable, appears to be both a fundamental requirement and an extremely difficult exercise, if only because neither its potential scope nor the statutory authorities that would lead the debate have been established at the outset. However, this impact is an essential factor at a time when new travel solutions that reduce the cost of mobility also bring with them the risk of unanticipated or unwanted regional developments, as has been the case at times of high levels of automobile use.

A Matter of Trust

Like all major disruptive innovations in mass transit modes, the transformation towards autonomous mobility requires a high level of trust that only the stakeholders closest to real life are likely to seek, even if they may not be able to guarantee it.

Trust relates above all to the safety of vehicles operating in traffic. The global impact of the Uber accident serves as a reminder that the public expects a much higher level of safety from an automated system than from a human being. Before a giant leap is made in human history, automated driving will first take several smaller steps, such as the implementation of driver assistance devices for personal vehicles on peri-urban highways and guidance devices to facilitate parking, driverless shuttle systems with fixed routes in environments where it is easy to get around, and reserved lanes on highways for automated public transportation vehicles.

Trust also relates to the use of personal data, which may be obtained legally or through hacking.

However, the matter of trust goes beyond these issues. It will also be necessary to demonstrate that the promises made can be fulfilled, in particular that the AV will contribute to a more pleasant urban environment, thanks to a reduction in traffic and congestion. This is still a long way off. Lessons learned from New York and several other major U.S. cities show that on-demand transportation service customers mainly comprise people who have reduced their use of public transport and eco-friendly modes or have abandoned them. Therefore, these services are contributing to an increase in traffic that these cities could certainly do without. The appealing concept of MaaS will only remain attractive and result in the anticipated benefits if powerful tools are used to discourage private car ownership and use. It's a power that "new world" players do not have in their arsenal, and would be reluctant to use even if they did.

Finally, trust will ultimately center around the type of region that will result from the implementation of these new forms of mobility. Admittedly, our current disillusionment with the automobile is at least partly down to its contribution to pollution and traffic jams, but it is also probably (and more fundamentally) due to the fact that its widespread use has transformed regions, to the point that discussions around the automobile have gradually shifted from opening up new possibilities to an obligation of use.

It is striking to note that despite the Internet, and regardless of (or perhaps because of) globalization and metropolization, citizens continue to expect a lot from their neighborhoods, value local mobility, show only modest enthusiasm for large metropolitan areas, and prefer cities to be on a "human scale." Many of citizens probably share the sentiment of Jean-Louis Guigou, when he was appointed the head of France's Interministerial Delegation of Land Planning and Regional Attractiveness (DATAR) in the late 1990s: "Globalization is imposed on us; the region depends on us."[248] The relative tolerance of citizens toward regulations on car use in cities shows that they want to move on from an era in which regions were shaped by the means of transport available, toward an approach where it is the regions themselves who are able to define the modes that can be used and their effects on the regional landscape.

Using the TRUST approach we argue for in this book, which we believe will benefit this shift in mobility as a whole, the question of how we want the regions we live, play, and work in to evolve should be at the top of the decision-making agenda, with the other factors (including regulations and systems and services) coming second.

There are many different facets to the region, such as its urban morphology (the structure of the city, which may be sparse or dense, coherent or incoherent), its environment (busy or quiet, conducive to encounters or unexpected events), its role as a place to live (where all city dwellers must feel safe, be treated with equal dignity, and have access to the resources they need to develop as people), and its collective contribution to society's ecological transition.

In this book, we wanted to draw attention to the risks that could be overlooked as a result of a fascination with the technology, which will always prove to be the greatest risk of all: it affects the nature of the spaces we inhabit, the relationships we build there, and the encounters those spaces enable.

---

248 Personal conversation with one of the authors.

The moment the autonomous vehicle leaves the laboratories, test tracks, and trial sites, it will become a "social object," just like the automobile before it. We can only trust, or indeed hope, that well-prepared local authorities and well-informed citizens will be able to civilize the self-driving car more quickly than they did its predecessor.

Glossary of Terms

Relating to Autonomous Vehicles and New Modes of Transportation

The authors would like to thank Guillaume Bresson and Jean-Laurent Franchineau of the VEDECOM Institute for their valuable input...and Wikipedia for being so universal.

Accelerometer A sensor that measures the longitudinal and lateral acceleration of a moving object. Often integrated into an inertial measurement unit.

Acceptability (adoption, desirability) The quality of being acceptable or tolerable. One of the criteria used to judge innovations and public policy, which, beyond being simply effective and efficient, must also be acceptable, or at least acceptable under certain circumstances. For example, the acceptability of a congestion charge may depend on how the proceeds are then used. Similarly, the acceptability of a self-driving car service may depend on the type of organization that offers the service. Acceptability does not imply adoption or desirability.

Access (Age of) A theory on the evolution of developed societies, primarily put forward by Jeremy Rifkin, which describes a shift in focus away from ownership to usage. In the context of this discussion, personal ownership of cars would be replaced by access to mobility services.

Actuator A device in a system that causes a physical action (e.g., a mechanical movement or an emission of heat or light) in response to a control signal (whether digital, electric, or hydraulic, etc.).

Advanced driver-assistance systems (ADAS) Driver-assistance systems range from those that simply provide information (e.g., warnings) and assistance (e.g., emergency braking without skidding or losing grip on the road) to those that perform specific driving tasks (e.g., keeping the vehicle in lane, dynamic speed regulation, and inter-vehicle distance control), which allow the driver to take back control at any time. These features are included in Levels 1 to 5 of the SAE J3016 "Levels of Driving Automation" standard.

Algorithm A finite and unambiguous sequence of operations or instructions used to solve a problem or generate a result. The term is mainly used in the field of computer science.

Approval Certification by a competent authority that an object is compliant with regulations, such as a highway code or technical standards. An approved vehicle may be used in the specific cases defined by the authority, and no other authorities may object to its use in those cases.

Artificial intelligence A set of theories and techniques using complex computer programs to simulate specific traits of human intelligence (e.g., reasoning and learning). It has applications in multiple fields, including artificial vision.

Automobile dependence A term popularized by Newman and Kenworthy in Cities and Automobile Dependence that refers to situations in which the ownership and use of cars is imposed on people, regardless of their wishes or preference. The term has a negative connotation due to its similarity to terminology used in situations of addiction.

Autonomous Refers to a device or system that can operate independently, without being connected to a power grid or information network (e.g., a bike or a sailboat).

Autonomous/self-driving/driverless (vehicle) A vehicle that can perform some or all driving tasks, in specific circumstances or at all times and with or without supervision, whether onboard or remote. "Self-driving" is a broader term than "driverless" which only refers to vehicles at the top level of the SAE J3016 scale.

Autonomy (of a person) An individual's ability to act independently, according to their own principles, abilities, interests, and beliefs.

Car-as-a-Service (CaaS) A vehicle used as a mobility service, which is not owned by its users.

Car sharing A system that enables a vehicle to be borrowed for a shorter period than that usually offered by car rental companies. We use the term "round-trip car sharing" when a vehicle is borrowed and returned to the same station, "one-way car sharing" when the vehicle is returned to another station, and "stationless car sharing" when the vehicle can be borrowed from and returned to any authorized parking point. The smartphone has become the main tool for car-sharing transactions (e.g., reservations, payments, and unlocking the car).

Carpooling Use of the same vehicle by several people to make a journey. This may be informal (carpooling between neighbors) or organized via a platform (such as BlaBlaCar) that matches offers and requests from strangers.

Co-modality A generic marketing term for a system that greatly facilitates multimodal and intermodal transportation.

Cone of Uncertainty The set of possible positions that an object could occupy within the next few seconds. It can be reduced to a single point (a fixed post) or extend over a large area (a running dog). The conical shape applies to objects with a particular set of dynamics. Other types of uncertainty can take on different forms depending on the various possible positions. For example, a GPS positioning system projected on a plane is an ellipse.

Connected (vehicle) A vehicle that exchanges real-time data with third parties (other vehicles or road infrastructure equipment), without the need for driver intervention.

Connected infrastructure All types of road infrastructure that are capable of communicating with vehicles and drivers, either digitally, using sound, by radio, or even visually.

Cyber (attack, crime, risk, security, hacking) Intrusion into a site, piece of software, hardware, or automated process, carried out by hackers to alter its intended behavior or steal its data. Cybersecurity aims to reduce or counter cyber risks.

Data

> ***Data:*** A basic piece of factual information, such as an observation or measurement. The increase in the amount of data being produced and stored in digital form today is exponential. It is estimated that the total volume of data doubles every other year. We all unintentionally contrib-

ute to this increase with our smartphones, credit cards, and Internet use, often referred to collectively as a "digital footprint." Some consider data to be the "oil of the twenty-first century."

Big Data: A situation linked to the coexistence of three factors: the large-scale collection of multi-source data in real time (e.g., texts, images, sounds, and digital data); a significant reduction in storage costs and an increase in computing speeds; and the development of data analysis techniques (e.g., data mining, data fusion, signal processing, neural networks, and machine learning) that are adapted to this context and geared towards action-taking and decision-making. Also referred to as the "three Vs," (volume, variety, velocity): the volume and variety of data and the speed of operations (e.g., storage, use, and analysis).

The circulation and use of personal data by major tech companies and their consolidated processing capacity are a continued cause for concern, despite the introduction of legal frameworks such as the General Data Protection Regulation by the European Union.

Data fusion: A set of methods for integrating data from heterogeneous and diverse sources (e.g., text, images, sounds, and digital data) in order to enrich each other through software processing.

Open data: Data that can be accessed and used by all, and that is distributed in a structured way according to a method and with an open license that specifies the terms and conditions of use. The move toward open public data is a way of satisfying the public's desire to take part in community life and to exercise the right to information, with public data being considered a common good and a condition for developers to offer services that add value to this data. In the latter case, conflicts may arise when the data is related to an activity of a competitive field, for example: real-time traffic data from a transport operator.

Disruption (disruptive innovation) A radical innovation that completely shakes up an established market. This concept stands in contrast to incremental innovation or the process of continuous improvement of products or services. The term has a positive connotation, unlike terms such as rupture, fracture, or even revolution. It first appeared in a 1995–1996 piece by Harvard Business School professor Clayton Christensen, which coincided with another by French advertising professional

Jean-Marie Dru, who offered a rather different meaning to the term.

Free and informed consent Consent given by a patient after being fully informed by a doctor of a treatment or operation. Consent collected by websites when the general terms of use are accepted. This type of consent is difficult to obtain from every person that may be exposed to an autonomous vehicle experiment.

Geolocation, GPS (Global Positioning System), GNSS (Global Navigation Satellite System) The term geolocation refers to any technique used to locate an object or assign an address to an object in a geographic information system. GPS and GNSS receivers use signals from a constellation of satellites to calculate the receiver's position. The level of accuracy when calculating a position is crucial for autonomous vehicles and relies on the technology used (e.g., dual frequency or base station).

Gyroscope A motion sensor that provides an angular velocity relative to an inertial reference (pitch or yaw roll axis as in aeronautics). Used in inertial measurement units.

Hybridization, hybrid vehicle

> *Hybridization:* A term borrowed from biology, referring to the process of creating a technology or service by borrowing characteristics from two different technologies or services. For example, bus rapid transit is ideally a bus route that is sufficiently protected and managed in such a way that its performance in terms of capacity, speed, and frequency is almost comparable to that of a subway line.
>
> *Hybrid vehicle:* A vehicle with an internal combustion engine and an electric motor.

Hype Media hype, frenzy, buzz. Gartner's Hype Cycle describes the change in attitudes toward new technologies, from overly optimistic expectations at the outset to more moderate expectations that adjust to the possibilities offered by technology.

Induced traffic, traffic evaporation Increase/decrease in traffic due to an increase/decrease in the capacity of the usable road space. This concept challenges the idea of a fixed demand for transportation.

Inertial measurement unit A navigation instrument that uses a combination of gyroscopes and accelerometers to estimate the position, direction of travel, and linear velocity of a moving object, relative to a starting point or checkpoint (e.g., the last known GPS position). It does not use any information external to the object, so it acts as a GPS relay when GPS signals are hidden or scrambled. Its measurements can be flawed and can lead to "drift" over time.

Inter-vehicle distance, safe following distance

> ***Inter-vehicle distance:*** The distance between a vehicle and the one in front of it.
>
> ***Safe following distance:*** The minimum distance between two vehicles that allows the vehicle behind to stop without colliding. Automated driving makes it possible to reduce the safe following distance needed by humans when driving, due to the faster reaction time of automated systems, and can therefore potentially increase the capacity of existing infrastructure.

Intermodality The use of several modes of transportation during a single journey. An intermodal hub is a place that is designed to facilitate the transition from one mode of transportation to another.

Internet of Things (IoT) The Internet of Things refers to the expansion of the Internet to include devices, objects, items, and places in the physical world. This Internet of connected things involves the exchange of information and data between real-world devices and the Internet. This includes connected vehicles.

Labeling The action of assigning a label to an entity. In the context of autonomous vehicles, it refers to the identification of an object (whether a living creature or inanimate object) within a virtual scene (such as the pixels of an image, for example).

Last mile An expression that refers to the final leg of a journey from the point where a means of transport stops (e.g., a bus stop or station) to the user's final destination. The "last mile problem" aims to mitigate the difficulties involved in this stage by providing complementary means of transport (e.g., a bicycle-sharing system or shuttles). Other considerations, such as public health (e.g., the health benefits of walking) and urban issues (e.g., the financial health of businesses along the route), are disregarded.

Lidar A device that emits a laser beam and receives reflected pulses (like radar), which allow it to determine its distance from an object.

MaaS (Mobility as a Service) An integrated system of mobility solutions (e.g., public transportation, self-service vehicles, and private hire services) that aims to make private ownership of vehicles unnecessary. The various components of the system can be accessed using the same device (usually a smartphone) and are combined in a single billing system.

Machine learning A method that enables a machine to improve its performance when carrying out a task, thanks to a learning phase based on collected data. This method is very useful in situations where it is difficult or impossible to formalize a priori rules for performing the desired task, for example, recognizing a form (a person, a cyclist, etc.) in an image, understanding natural language or handwriting, and so on. The development of machine learning has accelerated sharply since the early 2010s. Machine learning is a branch of artificial intelligence (AI).

Multimodality Describes a situation in which people routinely use a wide variety of transportation means and do not display a particular attachment to a preferred means.

Objects with agency An object that, by virtue of its very existence, generates and influences debate between people. As a case in point, talk of streetcar projects ended up dominating discussions on urban travel planning in the past, to the extent that the end goals of mobility for all and the smooth flow of traffic were sometimes overlooked.

Occupancy grid A representation of a vehicle's environment, which divides the environment into cells and positions the detected objects inside those cells. The result is a representation of the surroundings as detected by the vehicle's sensors, digitized and located on a mapped background, which can then be used to make predictions.

Onboard computer, onboard technology/system An onboard system is an electronic computer system (hardware and software) installed in an autonomous vehicle, which is designed to operate in real time, and which specializes in one or more specific tasks. The onboard computer performs less multi-tasking than a desktop computer, but also has greater constraints in terms of size, energy consumption, execution time, and above all, operational reliability.

Parking assistance, park assist function A system that makes the task of parking easier (rear movement radar system and ultrasonic sensors, for example) or performs it entirely under the driver's supervision.

Peer ratings A practice of allowing peers to review one another, developed by platforms as a way of building trust between people who do not know each other. Both providers and users are rated.

Personal mobility devices Devices such as hoverboards, rollerblades, skateboards, and scooters that offer an improvement to pedestrian performance. Such devices may be equipped with electric motors, giving them a range of dozens of miles. They can be used for leisure activities, but also for travel in the user's local area. The fact that they can be easily carried by users makes them an ideal option for journeys to and from public transport stops. Due to the newness of these devices, rules regulating their use on roads and sidewalks are not always clearly defined or remain yet to be formalized.

Platform A tool used as an intermediary between providers and users, and by extension, the companies that manage the tools (e.g., Airbnb, BlaBlaCar, and Uber). In the future, platforms could manage fleets of robo-taxis.

Radar (radio detection and ranging) A system that sends out electromagnetic waves to detect the presence of a target object and determine its position and speed using feedback signals (radar echoes).

Range (of an electric vehicle) The distance an electric vehicle with a fully charged battery can travel before it needs to recharge.

Reserved lanes, dedicated lanes Sections of roadways where parking is generally prohibited and that may only be used by designated vehicles (e.g., public transportation vehicles, emergency vehicles, bicycles and taxis in urban environments, buses and carpoolers on highways). These are often monitored by video surveillance.

Robo-taxi (or robot taxi) An autonomous vehicle that provides a taxi service via an app. The robo-taxi combines car sharing and taxi hailing (or private hire services). A number of industry insiders have pushed for robo-taxi systems as an alternative to private car use.

Self-driving See autonomous/self-driving/driverless (vehicle).

Sensor A device that detects a physical stimulus and outputs a signal in a usable format (usually digital).

Sharing (sharing economy, collaborative economy) Peer-to-peer human activity that aims to generate shared value or involves the use of pooled resources, space, or tools. These types of activity are often dependent on platforms that suggest and arrange matches between users.

Single-occupancy vehicle A term used to refer to the typical usage of a privately owned car, usually driven alone.

Smart highway, smart road, fifth-generation road A road equipped with sensors that allow it to monitor (e.g., detecting areas of wear and tear) and manage itself (e.g., by turning on street lighting), and equipped with telecommunications tools allowing it to communicate with the vehicles using it (known as V2I, or vehicle to infrastructure communication) and to facilitate communication between vehicles where needed (known as V2V, or vehicle-to-vehicle communication). It can also be sustainable (e.g., made of natural and/or recycled materials), energy-producing (e.g., using solar panel technology), and equipped with induction systems to recharge electric vehicles on the go.

Space consumption A measurement of the amount of urban space required to complete a trip, expressed in m^2 x hours. If a car occupies 10 m^2 when stationary, and 30 m^2 in urban traffic, a 30-minute trip with a single person on board followed by 2 hours of parking occupies: (0.5 x 30) + (2 x 10) = 35 m^2 x hours. A passenger on a bus for 30 minutes occupying 60 m^2 in traffic and that contains 30 people in total occupies: 0.5 x 60/30 = 1 m^2 x hours. Space consumption, along with vehicle noise and polluting emissions, are some of the criteria that help in defining urban planning for vehicles or systems.

Supervision A task performed by a human operator who is present, either in the autonomous vehicle (e.g., safety operator, emergency driver) or at a remote control station (in this case, one operator may supervise several vehicles). The task involves either identifying possible failures in the driving system or responding to notifications from this system, and then resolving any issues by taking control of the vehicle or providing instructions. In the case of remote supervision, the size of the data sets being transmitted means that highly efficient communication networks need to be put in place.

Takeover (time) The time required to switch from an automated driving situation to one where a human is driving and in full control of the vehicle.

TNC (Transportation Network Company) A company that matches passengers and drivers on demand and provides door-to-door transport. Unlike taxis, they can only be booked by prior reservation via the Internet or using a smartphone app (no hailing the streets), they do not have any distinctive visual markings, and their rates are unregulated.

Trusted third party A concept whose use has increased with the growth of e-commerce, electronic transactions, and the use of electronic signatures. A trusted third party is a natural or legal person, independent from the parties involved, who guarantees a transaction between them. By extension, the term "trusted third party" is used to refer to the natural or legal persons who are to be contacted in the event of a problem (e.g., a bus driver in the event of an altercation between passengers).

Update A new (often partial) version of a dataset or a piece of software, usually rolled out by download. Particularly important for continuously evolving software whose performance is improved from one version to the next.

Use case The description of a type of situation and environment where the autonomous capabilities of a vehicle can be experienced. For example: a daytime shuttle operated on an exclusive right of way on a suburban network, or a car on a suburban highway moving at a specified speed range.

V2I, V2V See smart highway.

White spot, dead zone, canyon An area where some telecommunication channels do not work (e.g., GPS in a tunnel).

Yann Leriche

Chief Executive Officer, Getlink

Yann Leriche is CEO of Getlink, a key player in mobility infrastructures, international transport and a leader in eco-responsible transport in Europe.

Leriche joined Getlink in summer of 2020, from Transdev North America where he was CEO since 2017. With extensive global experience leading large-scale public transportation businesses, he was responsible for bus, paratransit, streetcar, rail and light rail service contracts in more than 100 cities and communities in North America, with a workforce of 17,000 employees.

Known as a strategic forward thinker, Yann Leriche also led Transdev's global autonomous vehicle business and directed the Rouen Autonomous Lab, the first on-demand AV service on open roads in Europe. Shared ride AVs connect a tech business cluster to the city's robust public transit system.

Previously Leriche was Global Chief Performance Officer for Transdev, where he set and enforced standards for monitoring and improving operational performance globally. He also oversaw safety, IT, business development, marketing, and innovation. Before that, he was Deputy Chief Operating Officer of the North American Division in 2012 and 2013.

He also served as General Manager of Transamo, Transdev's consulting subsidiary, and as CEO of Transdev Germany's SZ (bus) and EurailCo (rail) divisions. Prior to that, he was Director of Public Transport in Reims, France and implemented a successful new light rail system that had a transformational effect on the city.

Leriche holds a Specialized Master's Degree in Financial and Fiscal Engineering from ESCP-Europe and additional degrees from College des Ingenieurs, Ecole Nationale des Ponts et Chaussees and Ecole Polytechnique.

Jean-Pierre Orfeuil Ph.D.

**Professor at Paris Institute of Urban Planning,
Chair of City on the Move Institute (IVM),
Academic Chair since 2000**

Jean-Pierre Orfeuil is a well-known and respected professor, researcher, and expert in mobility. Based in Paris, he has authored more than ten books on mobility, as well as hundreds of articles. He is an experienced and keen observer of what it takes to achieve urban transformation and the essential role that mobility and new mobility play in the quality of life of city dwellers. He is a frequent speaker at conferences and symposia focused on the importance of equitable access to quality mobility.

He graduated as a civil engineer from the École des Mines de Paris, a prestigious French engineering university. He holds a doctorate of statistics from the same university, and has been a professor at the Paris Institute of Urban Planning since 1998.

Jean-Pierre has also held the University Chair of the "City on the Move Institute" (Institut de la Ville en Mouvement - IVM) since 2000. IVM was designed to improve urban mobility and to act as a crossroads to bring together multi-disciplinary teams to execute unprecedented international projects and experiments around different axes of reflection in Asia, America, Europe and China. IVM is today an entity of the VEDECOM Institute.

He and the IVM advocate that mobility has become a fundamental human right because it determines access to other rights (housing, health, work, culture, education, etc). They focus on how the frequency, reliability and access to mobility/transport options play a major role in the quality of urban life.

Made in the USA
Monee, IL
04 November 2026

67384036R00108